ABOUT ISLAND PRESS

Island Press is the only nonprofit organization in the United States whose principal purpose is the publication of books on environmental issues and natural resource management. We provide solutions-oriented information to professionals, public officials, business and community leaders, and concerned citizens who are shaping responses to environmental problems.

In 1994, Island Press celebrates its tenth anniversary as the leading provider of timely and practical books that take a multidisciplinary approach to critical environmental concerns. Our growing list of titles reflects our commitment to bringing the best of an expanding body of literature to the environmental community throughout North America and the world.

Support for Island Press is provided by The Geraldine R. Dodge Foundation, The Energy Foundation, The Ford Foundation, The George Gund Foundation, William and Flora Hewlett Foundation, The James Irvine Foundation, The John D. and Catherine T. MacArthur Foundation, The Andrew W. Mellon Foundation, The Joyce Mertz-Gilmore Foundation, The New-Land Foundation, The Pew Charitable Trusts, The Rockefeller Brothers Fund, The Tides Foundation, Turner Foundation, Inc., The Rockefeller Philanthropic Collaborative, Inc., and individual donors.

EARTH
IN MIND

FOR MY FATHER AND MOTHER

EARTH IN MIND

On Education, Environment, and the Human Prospect

❖ ❖ ❖

DAVID W. ORR

ISLAND PRESS

WASHINGTON, DC • COVELO, CALIFORNIA

Library of Congress Cataloging-in-Publication Data

Orr, David W., 1944–
    Earth in mind : on education, environment, and the
human prospect / David W. Orr.
        p.   cm.
    Includes bibliographical references and index.
    ISBN (invalid) 1-55963-259-X (pbk.). —
ISBN 1-55963-294-1 (cloth)
    1. Environmental education—Philosophy.
2. Education—Philosophy.   I. Title.
GE70.O77   1994
363.7'0071—dc20   94-17546
                                              CIP

Printed on recycled, acid-free paper

Manufactured in the United States of America
10   9

CONTENTS

Part Four ❖ Destinations

ACKNOWLEDGMENTS

No PERSON is an island; I certainly am not. For their support, counsel, and friendship, I owe more to more people than I can possibly acknowledge. Some, however, deserve special thanks. First, I would like to thank my parents for their support and for their example of principled and purpose-filled lives. This book is dedicated to them with love and gratitude. I owe a special thank you to my wife, Elaine, who brings life, music, and joy to everything she does and to every life she touches, including my own. David Ehrenfeld, a renaissance man disguised as a biologist and serving as an editor, helped to greatly improve and clarify the essays here that originally appeared in *Conservation Biology*. I would like to thank friends who stimulated and sharpened many of the ideas found here, particularly Wendell Berry, Tony Cortese, Herman Daly, Kamyar Enshayan, George Foy, Wes Jackson, Gene Logsdon, Amory and Hunter Lovins, Jay McDaniel, Bill McDonough, Les Milbrath, John Powers, John and Nancy Todd, Steve Viederman, and George Woodwell. I would like to acknowledge all of those colleagues and friends at Oberlin College who make this a genial and stimulating place. None of these, however, should be held liable for my shortcomings revealed in the pages that follow. Finally, for permission to reprint articles that originally appeared elsewhere, I would like to thank the editors of *Annals of Earth*, *Conservation Biology*, *Ecological Economics*, *Holistic Education Review*, and Island Press.

INTRODUCTION

FROM newspapers, journal articles, and books, the following random facts crossed my desk within the past month:

- Male sperm counts worldwide have fallen by 50% since 1938, and no one knows exactly why.
- Human breast milk often contains more toxins than are permissible in milk sold by dairies.
- At death, human bodies often contain enough toxins and heavy metals to be classified as hazardous waste.
- Similarly toxic are the bodies of whales and dolphins washed up on the banks of the St. Lawrence River and the Atlantic shore.
- There has been a marked decline in fungi worldwide, and no one knows why.
- There has been a similar decline in populations of amphibians worldwide, even where the pH of rainfall is normal.
- Roughly 80% of European forests have been damaged by acid rain.
- U.S. industry releases some 11.4 billion tons of hazardous wastes to the environment each year.
- Ultraviolet radiation reaching the ground in Toronto is now increasing at 5% per year.

These facts only appear to be random. In truth, they are not random at all but part of a larger pattern that includes shopping malls and deforestation, glitzy suburbs and ozone holes, crowded freeways and climate change, overstocked supermarkets and soil erosion, a gross national product in excess of $5 trillion and superfund sites, and technological wonders and insensate violence. In reality there is no such thing as a "side effect" or an "externality." These things are threads of a whole cloth. The fact that we see them as disconnected events or fail to see them at all is, I believe, evidence of a considerable failure that we have yet to acknowledge as an educational failure. It is a failure to educate people to think broadly, to perceive systems and patterns, and to live as whole persons.

Much of the current debate about educational standards and reforms, however, is driven by the belief that we must prepare the young only to compete effectively in the global economy. That done, all will be well, or so it is assumed. But there are better reasons to reform education, which have to do with the rapid decline in the habitability of the earth. The kind of discipline-centric education that enabled us to industrialize the earth will not necessarily help us heal the damage caused by industrialization. Yale University historian Paul Kennedy (1993), after surveying the century ahead, reached broadly similar conclusions, calling for "nothing less than the re-education of humankind" (p. 331).

Yet we continue to educate the young for the most part as if there were no planetary emergency. It is widely assumed that environmental problems will be solved by technology of one sort or another. Better technology can certainly help, but the crisis is not first and foremost one of technology. Rather, it is a crisis within the minds that develop and use technology. The disordering of ecological systems and of the great biogeochemical cycles of the earth reflects a prior disorder in the thought, perception, imagination, intellectual priorities, and loyalties inherent in the industrial mind. Ultimately, then, the ecological crisis concerns how we think and the institutions that purport to shape and refine the capacity to think.

The essays in this book were written for different purposes and different audiences between 1990 and 1993. They are joined by the belief that the environmental crisis originates with the inability to think about ecological patterns, systems of causation, and the long-term effects of human actions. Eventually these are manifested as soil erosion, species extinction, deforestation, ugliness, pollution, social decay, injustice, and economic inefficiencies. In contrast, what can be called *ecological design intelligence* is the capacity to understand the ecological context in which humans live, to recognize limits, and to get the scale of things right. It is the ability to calibrate human purposes and natural constraints and do so with grace and economy. Ecological design intelligence is not just about things like technologies; it also has to do with the shape and dimension of our ideas and philosophies relative to the earth. At its heart ecological design intelligence is motivated by an ethical view of the world and our obligations to it. On occasion it requires the good sense and moral energy to say no to things otherwise possible and, for some, profitable. The surest signs of ecological design intelligence are collective achievements: healthy, durable, resilient, just, and prosperous communities.

I believe that educators must become students of the ecologically proficient mind and of the things that must be done to foster such minds. In time this will mean nothing less than the redesign of education itself.

SOURCES

Kennedy, P. 1993. *Preparing for the Twenty-First Century.* New York: Random House.

PART ONE

THE PROBLEM
OF EDUCATION

EDUCATION is not widely regarded as a problem, although the lack of it is. The conventional wisdom holds that all education is good, and the more of it one has, the better. The essays in Part one challenge this view from an ecological perspective. The truth is that without significant precautions, education can equip people merely to be more effective vandals of the earth. If one listens carefully, it may even be possible to hear the Creation groan every year in late May when another batch of smart, degree-holding, but ecologically illiterate, *Homo sapiens* who are eager to succeed are launched into the biosphere. The essays in Part one, accordingly, address the problem *of* education rather than problems *in* education. They are not a call to tinker with minutiae, but a call to deeper change.

❖

CHAPTER ONE

What Is Education For?

IF TODAY is a typical day on planet earth, we will lose 116 square miles of rain forest, or about an acre a second. We will lose another 72 square miles to encroaching deserts, the results of human mismanagement and overpopulation. We will lose 40 to 250 species, and no one knows whether the number is 40 or 250. Today the human population will increase by 250,000. And today we will add 2,700 tons of chlorofluoro-carbons and 15 million tons of carbon dioxide to the atmosphere. Tonight the earth will be a little hotter, its waters more acidic, and the fabric of life more threadbare. By year's end the numbers are staggering: The total loss of rain forest will equal an area the size of the state of Washington; expanding deserts will equal an area the size of the state of West Virginia; and the global population will have risen by more than 90,000,000. By the year 2000 perhaps as much as 20% of the life forms extant on the planet in the year 1900 will be extinct.

The truth is that many things on which our future health and prosperity depend are in dire jeopardy: climate stability, the resilience and productivity of natural systems, the beauty of the natural world, and biological diversity.

It is worth noting that this is not the work of ignorant people. Rather, it is largely the results of work by people with BAs, BSs, LLBs, MBAs, and PhDs. Elie Wiesel once made the same point, noting that the designers and perpetrators of Auschwitz, Dachau, and Buchenwald—the Holocaust—were the heirs of Kant and Goethe, widely thought to be the best educated people on earth. But their education did not serve as an adequate barrier to barbarity. What was wrong with their education? In Wiesel's (1990) words,

It emphasized theories instead of values, concepts rather than human beings, abstraction rather than consciousness, answers instead of questions, ideology and efficiency rather than conscience.

I believe that the same could be said of our education. Toward the natural world it too emphasizes theories, not values; abstraction rather than consciousness; neat answers instead of questions; and technical efficiency over conscience. It is a matter of no small consequence that the only people who have lived sustainably on the planet for any length of time could not read, or like the Amish do not make a fetish of reading. My point is simply that education is no guarantee of decency, prudence, or wisdom. More of the same kind of education will only compound our problems. This is not an argument for ignorance but rather a statement that the worth of education must now be measured against the standards of decency and human survival—the issues now looming so large before us in the twenty-first century. It is not education, but education of a certain kind, that will save us.

❖ Myth ❖

What went wrong with contemporary culture and education? We can find insight in literature, including Christopher Marlowe's portrayal of Faust who trades his soul for knowledge and power, Mary Shelley's Dr. Frankenstein who refuses to take responsibility for his creation, and Herman Melville's Captain Ahab who says "All my means are sane, my motive and my object mad." In these characters we encounter the essence of the modern drive to dominate nature.

Historically, Francis Bacon's proposed union between knowledge and power foreshadowed the contemporary alliance between government, business, and knowledge that has wrought so much mischief. Galileo's separation of the intellect foreshadowed the dominance of the analytical mind over that part given to creativity, humor, and wholeness. And in Descartes's epistemology, one finds the roots of the radical separation of self and object. Together these three laid the foundations for modern education, foundations that now are enshrined in myths that we have come to accept without question. Let me suggest six.

First, there is the myth that ignorance is a solvable problem. Ignorance is not a solvable problem; it is rather an inescapable part of the human condition. We cannot comprehend the world in its entirety. The advance of knowledge always carried with it the advance of some form of ignorance. For example, in 1929 the knowledge of what a substance like chlorofluorocarbons (CFCs) would do to the stratospheric ozone and climate stability was a piece of trivial ignorance as the compound had not yet been invented. But in 1930 after Thomas Midgely, Jr., discovered CFCs, what had been a piece of trivial ignorance became a critical life-threatening gap in human understanding of the biosphere. Not until the early 1970s did anyone think to ask "What does this substance do to what?" In 1986 we discovered that CFCs had created a hole in the ozone over the South Pole the size of the lower 48 U.S. states; by the early 1990s, CFCs had created a worldwide reduction of ozone. With the discovery of CFCs, knowledge increased, but like the circumference of an expanding circle, ignorance grew as well.

A second myth is that with enough knowledge and technology, we can, in the words of *Scientific American* (1989), "manage planet earth." Higher education has largely been shaped by the drive to extend human domination to its fullest. In this mission, human intelligence may have taken the wrong road. Nonetheless, managing the planet has a nice ring to it. It appeals to our fascination with digital readouts, computers, buttons, and dials. But the complexity of earth and its life systems can never be safely managed. The ecology of the top inch of topsoil is still largely unknown as is its relationship to the larger systems of the biosphere. What might be managed, however, is us: human desires, economies, politics, and communities. But our attention is caught by those things that avoid the hard choices implied by politics, morality, ethics, and common sense. It makes far better sense to reshape ourselves to fit a finite planet than to attempt to reshape the planet to fit our infinite wants.

A third myth is that knowledge, and by implication human goodness, is increasing. An information explosion, by which I mean a rapid increase of data, words, and paper is taking place. But this explosion should not be mistaken for an increase in knowledge and wisdom, which cannot be measured so easily. What can be said truthfully is that some knowledge is increasing while other kinds of knowledge are being lost. For example, David Ehrenfeld has pointed out that biology departments no longer hire faculty in such areas as systematics, taxonomy, or ornithology (personal communication). In other words, important knowledge is being lost because of the recent overemphasis on molecular biology and genetic engineering, which are more lucrative but not more important areas of inquiry. Despite all of our advances in some areas, we still do not have anything like the science of land health that Aldo Leopold called for a half-century ago.

It is not just knowledge in certain areas that we are losing but also vernacular knowledge, by which I mean the knowledge that people have of their places. According to Barry Lopez (1989),

> it is the chilling nature of modern society to find an ignorance of geography, local or national, as excusable as an ignorance of hand tools; and to find the commitment of people to their home places only momentarily entertaining, and finally naive.

> [I am] forced to the realization that something strange, if not dangerous, is afoot. Year by year the number of people with firsthand experience in the land dwindles. Rural populations continue to shift to the cities. . . . In the wake of this loss of personal and local knowledge, the knowledge from which a real geography is derived, the knowledge on which a country must ultimately stand, has come something hard to define but I think sinister and unsettling. (p. 55)

The modern university does not consider this kind of knowledge worth knowing except to record it as an oddity "folk culture." Instead, it conceives its mission as that of adding to what is called "the fund of human knowledge" through research. What can be said of research? Historian Page Smith (1990) has offered one answer:

> The vast majority of so-called research turned out in the modern university is essentially worthless. It does not result in any measurable benefit to anything or anybody. It does not push back those omnipresent 'frontiers of knowledge' so confidently evoked; it does not *in the main* result in greater health or happiness among the general populace or any particular segment of it. It is busywork on a vast, almost incomprehensible scale. It is dispiriting; it depresses the whole scholarly enterprise; and most important of all, it deprives the student of what he or she deserves—the thoughtful and considerate attention of a teacher deeply and unequivocally committed to teaching. (p. 7)

In the confusion of data with knowledge is a deeper mistake that learning will make us better people. But learning, as Loren Eiseley (1979) once said, is endless and "in itself . . . will never make us ethical men" (p. 284). Ultimately, it may be the knowledge of the good that is most threatened by all of our other advances. All things considered, it is possible that we are becoming more ignorant of the things we must know to live well and sustainably on the earth.

In thinking about the kinds of knowledge and the kinds of research that we will need to build a sustainable society, a distinction needs to be made between intelligence and cleverness. True intelligence is long range and aims toward wholeness. Cleverness is mostly short range and tends to break reality into bits and pieces. Cleverness is personified by the functionally rational technician armed with know-how and methods but without a clue about the higher ends technique should serve. The goal of education should be to connect intelligence with an emphasis on whole systems and the long range with cleverness, which involves being smart about details.

A fourth myth of higher education is that we can adequately restore that which we have dismantled. I am referring to the modern curriculum. We have fragmented the world into bits and pieces called disciplines and subdisciplines, hermetically sealed from other such disciplines. As a result, after 12 or 16 or 20 years of education, most students graduate without any broad, integrated sense of the unity of things. The consequences for their personhood and for the planet are large. For example, we routinely produce economists who lack the most rudimentary understanding of ecology or thermodynamics. This explains why our national accounting systems do not subtract the costs of biotic impoverishment, soil erosion, poisons in our air and water, and resource depletion from gross national product. We add the price of the sale of a bushel of wheat to the gross national product while forgetting to subtract the three bushels of topsoil lost to grow it. As a result of incomplete education, we have fooled ourselves into thinking that we are much richer than we are. The same point could be made about other disciplines and subdisciplines that have become hermetically sealed from life itself.

Fifth, there is a myth that the purpose of education is to give students the means for upward mobility and success. Thomas Merton (1985) once identified this as the "mass production of people literally unfit for anything except to take part in an elaborate and completely artificial charade" (p. 11). When asked to write about his own success, Merton responded by saying that "if it so happened that I had once written a best seller, this was a pure accident, due to inattention and naivete, and I would take very good care never to do the same again" (p. 11). His advice to students was to "be anything you like, be madmen, drunks, and bastards of every shape and form, but at all costs avoid one thing: success"

(p. 11). The plain fact is that the planet does not need more successful people. But it does desperately need more peacemakers, healers, restorers, storytellers, and lovers of every kind. It needs people who live well in their places. It needs people of moral courage willing to join the fight to make the world habitable and humane. And these qualities have little to do with success as our culture has defined it.

Finally, there is a myth that our culture represents the pinnacle of human achievement. This, of course, represents cultural arrogance of the worst sort and a gross misreading of history and anthropology. Recently, this view has taken the form that we won the Cold War. Communism failed because it produced too little at too high a cost. But capitalism has also failed because it produces too much, shares too little, also at too high a cost to our children and grandchildren. Communism failed as an ascetic morality. Capitalism has failed because it destroys morality altogether. This is not the happy world that any number of feckless advertisers and politicians describe. We have built a world of sybaritic wealth for a few and Calcuttan poverty for a growing underclass. At its worst, it is a world of crack on the streets, insensate violence, anomie, and the most desperate kind of poverty. The fact is that we live in a disintegrating culture. Ron Miller (1989) stated it this way:

> Our culture does not nourish that which is best or noblest in the human spirit. It does not cultivate vision, imagination, or aesthetic or spiritual sensitivity. It does not encourage gentleness, generosity, caring, or compassion. Increasingly in the late twentieth century, the economic-technocratic-statist worldview has become a monstrous destroyer of what is loving and life-affirming in the human soul. (p. 2)

❖ Rethinking Education ❖

Measured against the agenda of human survival, how might we rethink education? Let me suggest six principles.

First, all education is environmental education. By what is included or excluded, students are taught that they are part of or apart from the natural world. To teach economics, for example, without reference to the laws of thermodynamics or ecology is to teach a fundamentally important ecological lesson: that physics and ecology have nothing to do with the economy. It just happens to be dead wrong. The same is true throughout the curriculum.

A second principle comes from the Greek concept of Paideia. The goal of education is not mastery of subject matter but mastery of one's person. Subject matter is simply the tool. Much as one would use a hammer and a chisel to carve a block of marble, one uses ideas and knowledge to forge one's own personhood. For the most part we labor under a confusion of ends and means, thinking that the goal of education is to stuff all kinds of facts, techniques, methods, and information into the student's mind, regardless of how and with what effect it will be used. The Greeks knew better.

Third, I propose that knowledge carries with it the responsibility to see that it is well used in the world. The results of a great deal of contemporary research bear resemblance to those foreshadowed by Mary Shelley: monsters of technology and its byproducts for which no one takes responsibility or is even expected to take responsibility. Whose responsibility is Love Canal? Chernobyl? Ozone depletion? The *Exxon Valdez* oil spill? Each of these tragedies was possible because of knowledge created for which no one was ultimately responsible. This may finally come to be seen for what I think it is: a problem of scale. Knowledge of how to do vast and risky things has far outrun our ability to use it responsibly. Some of this knowledge cannot be used responsibly, safely, and to consistently good purposes.

Fourth, we cannot say that we know something until we understand the effects of this knowledge on real people and their communities. I grew up near Youngstown, Ohio, which was largely destroyed by corporate decisions to "disinvest" in the economy of the region. In this case MBA graduates, educated in the tools of leveraged buyouts, tax breaks, and capital mobility, have done what no invading army could do: They destroyed an American city with total impunity and did so on behalf of an ideology called the "bottom line." But the bottom line for society includes other costs: those of unemployment, crime, higher divorce rates, alcoholism, child abuse, lost savings, and wrecked lives. In this instance what was taught in the business schools and economics departments did not include the value of good communities or the human costs of a narrow destructive economic rationality that valued efficiency and economic abstractions above people and community (Lynd, 1982).

My fifth principle follows and is drawn from William Blake. It has to do with the importance of "minute particulars" and the power of examples over words. Students hear about global responsibility while being educated in institutions that often spend their budgets and invest their endowments in the most irresponsible things. The lessons being taught are those of hypocrisy and ultimately despair. Students learn, without anyone ever telling them, that they are helpless to overcome the frightening gap between ideals and reality. What is desperately needed are (a) faculty and administrators who provide role models of integrity, care, and thoughtfulness and (b) institutions capable of embodying ideals wholly and completely in all of their operations.

Finally, I propose that the way in which learning occurs is as important as the content of particular courses. Process is important for learning. Courses taught as lecture courses tend to induce passivity. Indoor classes create the illusion that learning only occurs inside four walls, isolated from what students call, without apparent irony, the "real world." Dissecting frogs in biology classes teaches lessons about nature that no one in polite company would verbally profess. Campus architecture is crystallized pedagogy that often reinforces passivity, monologue, domination, and artificiality. My point is simply that students are being taught in various and subtle ways beyond the overt content of courses.

❖ Reconstruction ❖

What can be done? Lots of things, beginning with the goal that no student should graduate from any educational institution without a basic comprehension of things like the following:

- the laws of thermodynamics,
- the basic principles of ecology,
- carrying capacity,
- energetics,
- least-cost, end-use analysis,
- limits of technology,
- appropriate scale,
- sustainable agriculture and forestry,
- steady-state economics, and
- environmental ethics.

I would add to this list of analytical and academic things, practical things necessary to the art of living well in a place: growing food; building shelter; using solar energy; and a knowledge of local soils, flora, fauna, and the local watershed. Collectively, these are the foundation for the capacity to distinguish between health and disease, development and growth, sufficient and efficient, optimum and maximum, and "should do" from "can do."

In Aldo Leopold's words, does the graduate know that "he is only a cog in an ecological mechanism? That if he will work with that mechanism his mental wealth and his material wealth can expand indefinitely? But that if he refuses to work with it, it will ultimately grind him to dust"? And Leopold asked, "If education does not teach us these things, then what is education for?" (p. 210).

SOURCES

Eiseley, L. 1979. *The Star Thrower*. New York: Harcourt Brace Jovanovich.

Leopold, A. 1966. *A Sand County Almanac*. New York: Ballantine. (Original work published 1949.)

Lopez, B. 1989, September. American Geographies. *Orion*.

Lynd, S. 1982. *The Fight Against Shutdowns*. San Pedro, CA: Singlejack Books.

Merton, T. 1985. *Love and Living*. New York: Harcourt Brace Jovanovich.

Miller, R. 1989, Spring. Editorial. *Holistic Education Review*.

Managing Planet Earth. 1989, Sept. *Scientific American, 261*, 3.

Smith, P. 1990. *Killing the Spirit*. New York: Viking.

Wiesel, E. 1990. Remarks before the Global Forum. Moscow.

CHAPTER TWO

The Dangers of Education

W E ARE currently preparing to launch yet another of our periodic national crusades to improve education. I am in favor of improving education, but what does it mean to improve education and what great ends will that improved education serve? The answer now offered from high places is that we must equip our youths to compete in the world economy. The great fear is that we will not be able to produce as many automobiles, VCRs, digital TVs, or supercomputers as the Japanese or Europeans. In contrast, I worry that we *will* compete all too effectively on an earth already seriously overstressed by the production of things economists count and too little production of things that are not easily countable such as well-loved children, good cities, healthy forests, stable climate, healthy rural communities, sustainable family farms, and diversity of all sorts. Many of the educational reforms now being proposed have little to do with the goals of personal wholeness, or the pursuit of truth and understanding, and even less to do with the great issues of how we might live within the limits of the earth. The reformers aim to produce people whose purposes and outlook are narrowly economic, not to educate citizens and certainly not "citizens of the biotic community."

The important facts of our time have more to do with too much economic activity of the wrong kind than they have to do with too little. Our means of livelihood are implicated everywhere in the sharp decline of the vital signs of the earth. Because of our fossil fuel–based economies and transportation systems, we are now conducting a risky and irreversible experiment with global climate. The same systems have badly damaged the ozone layer. The way we produce food and fiber is responsible for the loss of 24 billion tons of soil each year, the sharp decline in biological diversity, and the spread of deserts worldwide. The blind pursuit of national security has left a legacy of debt, toxicity, and radioactivity that will threaten the health and well-being of those purportedly defended for a long time to come. In addition, we continue to issue forth a stream of technologies and systems of technology that do not fit the ecological dimensions of the earth.

Most of this was not done by the unschooled. Rather it is the work of people who, in Gary Snyder's (1990) words,

> make unimaginably large sums of money, people impeccably groomed, excellently educated at the best universities—male and female alike—eating fine foods and reading classy literature, while orchestrating the investment and legislation that ruin the world. (p. 119)

Education, in other words, can be a dangerous thing. Accordingly, I intend to focus on the problem *of* education, not problems *in* education. It is time, I believe, for an educational "perestroika," by which I mean a general rethinking of the process and substance of education at all levels, beginning with the admission that much of what has gone wrong with the world is the result of education that alienates us from life in the name of human domination, fragments instead of unifies, overemphasizes success and careers, separates feeling from intellect and the practical from the theoretical, and unleashes on the world minds ignorant of their own ignorance. As a result, an increasing percentage of the human intelligence must attempt to undo a large part of what mere intellectual cleverness has done carelessly and greedily.

❖ Anticipations ❖

Most ancient civilizations knew what we have apparently forgotten: that knowledge is a fearful thing. To know the name of something was to hold power over it. Misused, that power would break the sacred order and wreak havoc. Ancient myths and legends are full of tales of people who believed that they were smarter than the gods and immune from divine punishment. But in whatever form, eating from the tree of knowledge meant banishment from one garden or another. In the modern world this Janus-like quality of knowledge has been forgotten. Descartes, for example, reached the conclusion that "the more I sought to inform myself, the more I realized how ignorant I was." Instead of taking this as a proper conclusion of a good education, Descartes set about to find certain truths through a process of radical skepticism. Francis Bacon went even further, to propose an alliance between science and power, which reached fruition in the Manhattan Project and the first atomic bomb.

There were warnings, however. Displaced tribal peoples commonly regarded Europeans as crazy. In 1744, for example, the Chiefs of the Six Nations declined an offer to send their sons to the College of William and Mary in these words:

> Several of our young people were formerly brought up at the colleges of the northern provinces: they were instructed in your sciences; but when they came back to us, they were bad runners, ignorant of every means of living in the woods . . . neither fit [to be] hunters, warriors, nor counsellors, they were totally good for nothing. (McLuhan, 1971, p. 57)

Native Americans detected the lack of connectedness and rootedness that Europeans, with all of their advancements, could not see in themselves. European education incapacitated whites in ways visible only through the eyes of people whose minds still participated in the creation and for whom the created order was still enchanted. In other words, European minds were not prepared for the encounter with wilderness nor were they prepared to understand those who could live in it. One had to step out of the dominant Eurocentrism and see things from the outside looking in. A century later Ralph Waldo Emerson was moving toward a similar conclusion:

> We are shut up in schools and college recitation rooms for ten or fifteen years, and come out at least with a bellyful of words and do not know a thing. We cannot use our hands, or our legs, or our eyes or our arms. We do not know an edible root in the woods. We cannot tell our course by the stars, nor the hour of the day by the sun. (p. 136)

These and other warnings were forebodings of a much more serious problem that would gain momentum in the century to come. I think this becomes clearer in a comparison of two prominent but contrary figures of the middle years of the twentieth century.

One, Albert Speer, was born in Germany in 1905 to a well-to-do upper-middle-class family. His father was one of the busiest architects in the booming industrial city of Mannheim. Speer attended a distinguished private school and later various institutes of technology in Karlsruhe,

Munich, and Berlin. At the age of 23, Speer became a licensed architect. He is not known to us for his architecture, however, but for his organizational genius as Hitler's Minister of Armaments. In that role he kept World War II going far longer than it otherwise would have by keeping German arms production rising under the onslaught of Allied bombing until the final months. For his part in extending the war and for using slave labor to do so, Speer was condemned by the Nuremburg Tribunal to serve 20 years at Spandau Prison.

I think Speer's teachers and professors should share some of the blame. For example, in his memoirs Speer (1970) described his education as apolitical:

> [Our education] impressed upon us that the distribution of power in society and the traditional authorities were part of the God-given order of things. . . . It never occurred to us to doubt the order of things. (p. 8)

The result was a "generation without defenses" for the seductions of Hitler and the new technologies of political persuasion. The best educated nation in Europe had no civic education when it most needed it. Speer was not appreciably different from millions of others swept along by the current of Nazism.

The purge of June 30, 1934, was a moral turning point after which Speer silenced all doubts about his role in the Nazi hierarchy:

> I saw a large pool of dried blood on the floor. There on June 30 Herbert Von Bose, one of Papen's assistants, had been shot. I looked away and from then on avoided the room. But the incident did not affect me more deeply than that. (p. 53)

Speer had found his Mephistopheles:

> After years of frustrated efforts I was wild to accomplish things— and twenty-eight years old. For the commission to do a great building, I would have sold my soul like Faust. Now I had found my Mephistopheles. He seemed no less engaging than Goethe's. (p. 31)

In looking back over his life near its end, Speer made the following comment:

> My moral failure is not a matter of this item and that; it resides in my active association with the whole course of events. I had participated in a war which, as we of the intimate circle should never have

doubted, was aimed at world dominion. What is more, by my abilities and my energies I had prolonged that war by many months. . . . Dazzled by the possibilities of technology, I devoted crucial years of my life to serving it. But in the end my feelings about it are highly skeptical. (pp. 523–524)

Finally, in what certainly would be among the most plaintive lines penned by any leading figure of the twentieth century, Speer wrote, "The tears I shed are for myself as well as for my victims, for the man I could have been but was not, for a conscience I so easily destroyed."

If Speer and the years between 1933 and 1945 seem remote from the issues of the late twentieth century, one has only to change the names to see a relationship. Instead of World War II, think of the war being waged against nature. Instead of the Holocaust think of the biological holocaust now under way in which perhaps 20% of the life forms on the planet in the year 1900 will have disappeared by the early years of the next century. Instead of the fanaticism of the 1000-year Reich, think of the fanaticism inherent in the belief that economies have no limits and can grow forever. Speer's upbringing and formal education provided neither the wherewithal to think about the big issues of his time nor the good sense to call these by their right names. I do not think for a moment that this kind of education ended in 1945. It remains the predominant mode of education almost everywhere in an age that still regards economic growth as the highest goal.

Like Speer, Aldo Leopold was middle-class, the son of a prosperous furniture manufacturer (in Burlington, Iowa) and had all the advantages of good upbringing (Meine, 1988). Leopold's lifelong study of nature began as a boy in the nearby marshes along the Mississippi River. His formal education at Lawrence Academy in New Jersey and at Yale University were, I think, rather incidental to his self-education, which consisted of long walks over the nearby countryside. Leopold was an outdoorsman who, over a lifetime of rambling, developed the ability to observe in nature what others could not see. He was a keen student of nature, and it was this capacity that makes Leopold interesting and important to us. Leopold grew from a rather conventional resource manager employed by the U.S. Forest Service to become a scientist and philosopher who asked questions about the proper human role in nature that no one else bothered to ask. This progression led him to discard the idea of human dominance and to propose more radical ideas on the basis of our citizenship in the natural order.

Where Speer had seen human blood on the floor and turned away, Leopold described a different kind of turning point that took place on a rimrock overlooking a river in the Gila Wilderness in 1922. Leopold and his companions spotted a she-wolf and cubs along the bank and opened fire:

> We reached the old wolf in time to watch a fierce green fire dying in her eyes. I realized then, and have known ever since, that there was something new to me in those eyes—something known only to her and to the mountain. I was young then and full of trigger itch; I thought that because fewer wolves meant more deer, then no wolves would mean a hunters' paradise. But after seeing the green fire die, I sensed that neither the wolf nor the mountain agreed with such a view. (Leopold, 1966, pp. 137–139)

The rest of Leopold's life was an extended meditation on that fierce green fire, how mountains think, and what both meant for humans.

Where Speer regarded himself as apolitical, Leopold (1966) regarded "biological education as a means of building citizens" (p. 208). Instead of possessing a deep naivete about science, Leopold (1991) was scientific about science as few have ever been:

> We are not scientists. We disqualify ourselves at the outset by professing loyalty to and affection for a thing: wildlife. A scientist in the old sense may have no loyalties except to abstractions, no affections except for his own kind. . . . The definitions of science written by, let us say, the National Academy, deal almost exclusively with the creation and exercise of power. But what about the creation and the exercise of wonder or respect for workmanship in nature? (p. 276)

Where Speer's (1970) approach to nature was sentimental and escapist (to escape "the demands of a world growing increasingly complicated"), Leopold's (1966) was hardheaded and practical:

> The cultural value of wilderness boils down in the last analysis, to a question of intellectual humility. The shallow minded modern who has lost his rootage in the land assumes that he has already discovered what is important; it is such who prate of empires, political or economic, that will last a thousand years. (p. 279)

Where Speer had to learn his ethics in 20 years of confinement after the damage was done, Leopold learned his over a lifetime and laid the basis for an ecologically solvent land ethic. And where Speer's education made him immune to seeing or feeling tragedy unfolding around him, Leopold (1966) wrote the following:

> One of the penalties of an ecological education is that one lives alone in a world of wounds. Much of the damage inflicted on land is quite invisible to laymen. An ecologist must either harden his shell and make believe that the consequences of science are none of his business, or he must be the doctor who sees the marks of death in a community that believes itself to be well and does not want to be told otherwise. (p. 197)

After Speer and the Nazis, it has taken decades to undo the damage that could be undone. After Aldo Leopold, in contrast, it will take decades to fully grasp what he meant by a "land ethic" and considerably longer to make it a reality.

❖ Dangers ❖

From the lives of Speer and Leopold, what can be said about the dangers of formal education or schooling? This first and overriding danger is that it will encourage young people to find careers before they find a decent calling. A career is a job, a way to earn one's keep, a way to build a long resume, a ticket to somewhere else. For upwardly mobile professionals, a career is too often a way to support a "lifestyle" by which one takes more than one gives back. In contrast, a calling has to do with one's larger purpose, personhood, deepest values, and the gift one wishes to give the world. A calling is about the use one makes of a career. A career is about specific aptitudes; a calling is about purpose. A career is planned with the help of "career development" specialists. A calling comes out of an inner conversation. A career can always be found in a calling, but a calling cannot easily be found in a career. The difference is roughly like deciding to which end of the cart to attach the horse. Speer's problem was not a deficiency of mathematical skills, or reading ability, or computing ability, or logic narrowly conceived. I imagine that he would have done well on the Scholastic Aptitude Test or the Graduate Record Exam. His problem was simply that he had no calling that could bridle and channel his ambition. He simply wanted to "succeed," doing whatever it took. He was, as he said, "wild to accomplish," and ambition disconnected the alarm bells that should have sounded long before he saw blood on the floor in 1934. Speer was a careerist with no calling.

Leopold, on the other hand, found his calling as a boy in the marshes around Burlington, Iowa, and followed it wherever it took him. In time it took him a long way. From his boyhood interest in birds, he went on in adult life to initiate the field of game management, to organize the Wilderness Society, to work actively on behalf of conservation throughout his lifetime, and to lay the groundwork for the field of environmental ethics, while still finding time to be a good teacher and a good father. There is a consistency and harmony to Leopold's life rather like a pilgrim following a vision.

A second danger of formal schooling is that it will imprint a disciplinary template onto impressionable minds and with it the belief that the world really is as disconnected as the divisions, disciplines, and subdisciplines of the typical curriculum. Students come to believe that there is such a thing as politics separate from ecology or that economics has nothing to do with physics. Yet, the world is not this way, and except for the temporary convenience of analysis, it cannot be broken into disciplines and specializations without doing serious harm to the world and to the minds and lives of people who believe that it can be. We often forget to tell students that the convenience was temporary, and more seriously, we fail to show how things can be made whole again. One result is that students graduate without knowing how to think in whole systems, how to find connections, how to ask big questions, and how to separate the trivial from the important. Now more than ever, however, we need people who think broadly and who understand systems, connections, patterns, and root causes.

This is an unlikely outcome of education conceived as the propagation of technical intelligence alone. Speer in his Nazi years was a technician and a good one. His formal schooling gave him the tools that could be used by the Third Reich but not the sense to ask why and not the humanity necessary to recognize the face of barbarity when he saw it. Leopold, in contrast, began his career as something of a technician, but outgrew it. A Sand County Almanac, written shortly before his death, was a nearly perfect blend of science, natural history, and philosophy.

Third, there is the danger that education will damage the sense of wonder—the sheer joy in the created world—that is part of our original equipment at birth. It does this in various ways: by reducing learning to routines and memorization, by excess abstractions divorced from lived experience, by boring curriculum, by humiliation, by too many rules, by overstressing grades, by too much television and too many computers, by too much indoor learning, and mostly by deadening the feelings from which wonder grows. As our sense of wonder in nature diminishes, so too does our sense of the sacred, our pleasure in the created world, and the impulse behind a great deal of our best thinking. Where it is kept intact and growing, teachers need not worry about whether students learn reading, writing, and arithmetic.

In a small book titled *The Sense of Wonder*, Rachel Carson (1984) wrote that "it is not half so important to *know* as to *feel*" (p. 45). Feelings, she wrote, begin early in life in the exploration of nature, generally with the companionship of an adult. The sense of wonder is rooted in the trust that the world is, on balance, a friendly place full of interesting life "beyond the boundaries of human existence" (p. 88). The sense of wonder that Carson describes is not equivalent to a good science education, although in principle I see no reason why the two cannot be made compatible. I do not believe that wonder can be taught as "Wonder 101." If Carson is right, it can only be felt, and those early feelings must be encouraged, supported, and legitimized by a caring and knowledgeable adult. My hunch is that the sense of wonder is fragile; once crushed, it rarely blossoms again but is replaced by varying shades of cynicism and disappointment in the world.

I know of no measures for wonder, but I think Speer lost his early on. His relation to nature prior to 1933 was, by his testimony, romantic and escapist. Thereafter, he mentioned it no more. To Speer, the adult, the natural world was not particularly wondrous, nor was it a source of insight, pleasure, or perspective. His orientation toward life, like that of the Nazi hierarchy, was necrophilic. Leopold, on the contrary, was a life-long student of nature in the wild. By all accounts he was a remarkably astute observer of land, which explains a great deal of his utter sanity and clarity of mind. Leopold's intellectual and spiritual anchor was not forged in a laboratory or a library but in time spent in the wild and in his later years in a rundown farm he purchased that the family called "the shack."

❖ Conclusion ❖

What are the dangers of education? There are three that are particularly consequential for the way we live on the earth: (1) that formal education will cause students to worry about how to make a living before they know who they are, (2) that it will render students narrow technicians who are morally sterile, and (3) that it will deaden their sense of wonder for the created world. Of course education cannot do these things alone. It requires indifferent or absentee parents, shopping malls, television–MTV–Nintendo, a culture aimed at the lowest common denominator, and de-placed people who do not know the very ground beneath their feet. Schooling is only an accomplice in a larger process of cultural decline. Yet, no other institution is better able to reverse that decline. The answer, then, is not to abolish or diminish formal education but rather to change it.

SOURCES

Carson, R. 1984. *The Sense of Wonder*. New York: Harper.

Emerson, R. W. 1972. *Selections from Ralph Waldo Emerson*. S. E. Whicher, ed. Boston: Houghton-Mifflin. (Original work published 1839.)

Leopold, A. 1966. *A Sand County Almanac*. New York: Ballantine. (Original work published 1949.)

Leopold, A. 1991. *The River of the Mother of God and Other Essays by Aldo Leopold*. S. Flader and J. B. Callicott, eds. Madison: University of Wisconsin Press. (Original work published 1941.)

McLuhan, T. C. 1971. *Touch the Earth*. New York: Simon & Schuster.

Meine, C. 1988. *Aldo Leopold: His Life and Work*. Madison: University of Wisconsin Press.

Snyder, G. 1990. *The Practice of the Wild*. San Francisco: North Point Press.

Speer, A. 1970. *Inside the Third Reich*. Boston: Houghton Mifflin.

CHAPTER THREE

The Problem of Education

AFTER due reflection on the state of education in his time, H. L. Mencken concluded that significant improvement required only that the schools be burned to the ground and all of the professorate be hanged. For better or worse, the suggestion was ignored. Made today, however, it might find a more receptive public ready to purchase the gasoline and rope. Americans, united on little else, seem of one mind in believing that, K through PhD, the educational system is too expensive, too cumbersome, and not, on the whole, very effective. It needs, they believe, radical reform. They are, however, divided on how to go about it.

Both sides of the debate, nonetheless, agree on the basic aims and purposes of education, which are to equip our nation with a "world-class" labor force, first, to compete more favorably in the global economy and, second, to provide each individual with the means for maximum upward mobility. On these, the purposes of education both higher and lower, there is great repose.

There are, nonetheless, better reasons to rethink education that have to do with the issues of human survival, which will dominate the world of the twenty-first century. Those now being educated will have to do what we, the present generation, have been unable or unwilling to do: stabilize world population; stabilize and then reduce the emission of greenhouse gases, which threaten to change the climate, perhaps disastrously; protect biological diversity; reverse the destruction of forests everywhere; and conserve soils. They must learn how to use energy and materials with great efficiency. They must learn how to utilize solar energy in all of its forms. They must rebuild the economy in order to eliminate waste and pollution. They must learn how to manage renewable resources for the long run. They must begin the great work of repairing, as much as possible, the damage done to the earth in the past 200 years of industrialization. And they must do all of this while they reduce worsening social and racial inequities. No generation has ever faced a more daunting agenda.

For the most part, however, we are still educating the young as if there were no planetary emergency. Remove computers and a scattering of courses and programs throughout the catalog, and the curriculum of the 1990s looks a lot like that of the 1950s. The crisis we face is first and foremost one of mind, perception, and values; hence, it is a challenge to those institutions presuming to shape minds, perceptions, and values. It is an educational challenge. More of the same kind of education can only make things worse. This is not an argument against education but rather an argument for the kind of education that prepares people for lives and livelihoods suited to a planet with a biosphere that operates by the laws of ecology and thermodynamics.

The skills, aptitudes, and attitudes necessary to industrialize the earth, however, are not necessarily the same as those that will be needed to heal the earth or to build durable economies and good communities. Resolution of the great ecological challenges of the next century will require us to reconsider the substance, process, and purpose of education at all levels and to do so, in the words of Yale University historian Jaroslav Pelikan (1992), "with an intensity and ingenuity matching that shown by previous generations in obeying the command to have dominion over the planet (p. 21). But Pelikan (1992) himself doubts whether the university "has the capacity to meet a crisis that is not only ecological and technological, but ultimately educational and moral" (pp. 21–22). Why should this be so? Why should those institutions charged with the task of preparing the young for the challenges of life be so slow to recognize and act on the major challenges of the coming century?

A clue can be found in a recent book by Derek Bok (1990), a former president of Harvard University, who wrote,

> Our universities excel in pursuing the easier opportunities where established academic and social priorities coincide. On the other hand, when social needs are not clearly recognized and backed by adequate financial support, higher education has often failed to respond as effectively as it might, even to some of the most important challenges facing America. Armed with the security of tenure and the time to study the world with care, professors would appear to have a unique opportunity to act as society's scouts to signal impending problems. . . . Yet rarely have members of the academy succeeded in

discovering emerging issues and bringing them vividly to the attention of the public. What Rachel Carson did for risks to the environment, Ralph Nader for consumer protection, Michael Harrington for problems of poverty, Betty Friedan for women's rights, they did as independent critics, not as members of a faculty. (p.105)

This observation, appearing on page 105 of Bok's book, is not mentioned thereafter. It should have been on page 1 and would have provided the subject for a better book. Had Bok gone further, he might have been led to ask whether the same charge of lethargy might be made against those presuming to lead American education. Bok might then have been led to rethink old and unquestioned assumptions about liberal education. For example, John Henry Newman (1982), in his classic *The Idea of a University*, drew a distinction between practical and liberal learning that has influenced education from his time to our own. Liberal knowledge, according to Newman, "refuses to be informed by any end, or absorbed into any art" (p. 81); knowledge is liberal if "nothing accrues of consequence beyond the using" (p. 82). Furthermore, Newman stated that "liberal education and liberal pursuits are exercises of mind, of reason, of reflection" (p. 80). All else he regarded as practical learning, which Newman believed has no place in the liberal arts. To this day, Newman's distinction between practical and liberal knowledge is seldom transgressed in liberal arts institutions. Is it any wonder that faculty, mindful of the penalties for transgressions of one sort or another, do not often deal boldly with the kinds of issues that Bok describes? I do not wish to take faculty off the hook, but I would like to note that educational institutions, more often than not, reward indoor thinking, careerism, and safe conformity to prevailing standards. Educational institutions are not widely known for encouraging boat rockers, and I seriously doubt that Bok's own institution would have awarded tenure to Rachel Carson, Ralph Nader, or Michael Harrington.

Harvard philosopher and mathematician Alfred North Whitehead had a different view of the liberal arts. "The mediocrity of the learned world," he wrote in 1929, could be traced to its "exclusive association of learning with book-learning" (Whitehead 1967, p. 51). Whitehead went on to say that real education requires "first-hand knowledge," by which he meant an intimate connection between the mind and "material creative activity." Others, such as John Dewey and J. Glenn Gray, reached similar conclusions. "Liberal education," Gray (1984) wrote, "is least dependent on formal instruction. It can be pursued in the kitchen, the workshop, on the ranch or farm . . . where we learn wholeness in response to others" (p. 81). A genuinely liberal education, in other words, ought to be liberally conducted, aiming to develop the full range of human capacities. And institutions dedicated to the liberal arts ought to be more than simply agglomerations of specializations.

Had Bok proceeded further he would have had to address the loss of moral vision throughout education at all levels. In ecologist Stan Rowe's (1990) words the university has

> shaped itself to an industrial ideal—the knowledge factory. Now it is overloaded and top-heavy with expertness and information. It has become a know-how institution when it ought to be a know-why institution. Its goal should be deliverance from the crushing weight of unevaluated facts, from bare-bones cognition or ignorant knowledge: knowing in fragments, knowing without direction, knowing without commitment. (p. 129)

Many years ago William James (1987) saw this coming and feared that the university might one day develop into a "tyrannical Machine with unforeseen powers of exclusion and corruption" (p. 113). We are moving along that road and should ask why this has come about and what can be done to reverse course.

One source of the corruption is the marriage between the academy and the worlds of power and commerce. It was a marriage first proposed by Francis Bacon, but not fully consummated until the later years of the twentieth century. But marriage, implying affection and mutual consent, is perhaps not an accurate metaphor. This is instead a cash relationship, which began with a defense contract here and a research project there. At present more than a few university departments still work as adjuncts of the Pentagon and even more as adjuncts of industry in the hope of reaping billions of dollars in fields such as genetic engineering, nanotechnologies, agribusiness, and computer science. Even where this is not true, it is difficult to escape the conclusion that much of what passes for research, as historian Page Smith (1990) wrote, is "essentially worthless . . . busywork on a vast almost incomprehensible scale" (p. 7).

Behind the glossy facade of the modern academy there is often a vacuum of purpose waiting to be filled by whomever and whatever. For example, the College of Agriculture at a nearby land-grant university of note claims to be helping "position farmers for the future." But when asked what farming would be like in the twenty-first century, the Dean of the College replied by saying, "I don't know." When asked, "How can you [then] position yourself for it?" the Dean replied, "We have to try as best we can to plan ahead" (Logsdon, 1994, p. 74). This reminds me of the old joke in which the airline pilot reports to the passengers that he has good news and bad news. The good news is that the flight is ahead of schedule. The bad news? "We're lost." And in a time of eroding soils and declining rural communities, "turf grass management" is the hot new item at the college of agriculture.

Finally, had Bok so chosen, he would have been led to question how we define intelligence and what that might imply for our larger prospects. At the heart of our pedagogy and curriculum is a fateful confusion of cleverness with intelligence. Cleverness, as I understand it, tends to fragment things and to focus on the short run. The epitome of cleverness is the specialist whose intellect and person have been shaped by the demands of a single function, what Nietzsche once called an "inverted cripple." Ecological intelligence, on the other hand, requires a broader view of the world and a long-term perspective. Cleverness can be adequately computed by the Scholastic Aptitude Test and the Graduate Record Exam, but intelligence is not so easily measured. In time I think we will come to see that true intelligence tends to be integrative and often works slowly while one is mulling things over.

The modern fetish with smartness is no accident. The highly specialized, narrowly focused intellect fits the demands of instrumental rationality built into the industrial economy, and for reasons described by Brooks Adams (cited in Smith, 1984) 80 years ago,

> capital has preferred the specialized mind and that not of the highest quality, since it has found it profitable to set quantity before quality to the limit the market will endure. Capitalists have never insisted upon raising an educational standard save in science and mechanics, and the relative overstimulation of the scientific mind has now become an actual menace to order. (p. 116)

The demands of building good communities within a sustainable society in a just world order will require more than the specialized, one-dimensional mind and more than instrumental cleverness.

❖ The Task ❖

Looking ahead to the twenty-first century, I see the task of educating minds capable of building a sustainable world order as requiring more comprehensive and ecologically solvent standards for truth. The architects of the modern worldview, notably Galileo and Descartes, assumed that those things that could be weighed, measured, and counted were more true than those that could not be quantified. If it could not be counted, in other words, it did not count. Cartesian philosophy was full of potential ecological mischief, a potential that has become reality. Descartes's philosophy separated man from nature, stripped all intrinsic value from nature, and then proceeded to divide mind and body. Descartes was, at heart, an engineer, and his legacy to the environment of our time is the cold passion to remake the world as if we were merely remodeling a machine. Feelings and intuition were tossed out, as were those fuzzy, qualitative parts of reality, such as aesthetic appreciation, loyalty, friendship, sentiment, empathy, and charity. Descartes's assumptions were neither as simple nor as inconsequential as they might have appeared in his lifetime (1596–1650).

If saving species and environments is our aim, we will need a broader conception of science and a more inclusive rationality that joins empirical knowledge with the same emotions that make us love and sometimes fight. Philosopher Karl Polanyi (1958) described this as "personal knowledge," by which he meant knowledge that calls forth a wider range of human perceptions, feelings, and intellectual powers than those presumed to be narrowly "objective." Personal knowledge, according to Polanyi,

> is not made but discovered. . . . It commits us, passionately and far beyond our comprehension, to a vision of reality. Of this responsibility we cannot divest ourselves by setting up objective criteria of verifiability—or falsifiability, or testability. . . . For we live in it as in the garment of our own skin. Like love, to which it is akin, this commitment is a 'shirt of flame', blazing with passion and, also like love, consumed by devotion to a universal demand. Such is the true sense of objectivity in science . . . (p. 64)

Cartesian science rejects passion and personality but ironically can escape neither. Passion and personality are embedded in all knowledge, including the most ascetic scientific knowledge driven by the passion for objectivity. Descartes and his heirs simply had it wrong. There is no way to separate feeling from knowledge. There is no way to separate object from subject. There is no good way and no good reason to separate mind or body from its ecological and emotional context. And some persons, with good evidence, are coming to suspect that intelligence is not a human monopoly at all (Griffin, 1992). Science without passion and love can give us no reason to appreciate the sunset, nor can it give us any purely objective reason to value life. These must come from deeper sources.

Second, we will have to challenge the hubris buried in the hidden curriculum that says that human domination of nature is good; that the growth of economy is natural; that all knowledge, regardless of its consequences, is equally valuable; and that material progress is our right. As a result we suffer a kind of cultural immune deficiency anemia that renders us unable to resist the seductions of technology, convenience, and short-term gain. In this perspective, the ecological crisis is a test of our loyalties and of our deeper affinities for life: what Harvard biologist Edward O. Wilson (1984) calls "biophilia."

Third, the modern curriculum teaches little about citizenship and responsibilities and a great deal about individualism and rights. The ecological emergency, however, can be resolved only if enough people come to hold a bigger idea of what it means to be a citizen. This will have to be carefully taught at all levels of education, but a pervasive cynicism about our higher potentials and collective abilities now works against us. Even my most idealistic students often confuse self-interest with selfishness, a view that describes both Mother Teresa and Donald Trump as self-maximizers, each merely doing "her thing" or "his thing." This is not just a social and political problem. The ecological emergency is about the failure to comprehend our citizenship in the biotic community. From the modern perspective we cannot see clearly how utterly dependent we are on the "services of nature" and on the wider community of life. Our political language gives little hint of this dependence. As it is now used, the word *patriotism*, for example, is devoid of ecological content. However, it must come to include the use one makes of land, forests, air, water, and wildlife. To abuse natural resources, to erode soils, to destroy natural diversity, to waste, to take more than one's fair share, to fail to replenish what has been used must someday come to be regarded as unpatriotic and wrong. And "politics" once again must come to mean, in Vaclav Havel's (1992) words, "serving the community and serving those who will come after us" (p. 6).

Fourth, there is a widespread, and mostly unquestioned, assumption that our future is one of constantly evolving technology and that this is always and everywhere a good thing. Those who question this faith are dismissed as Luddites by people who, as far as I can tell, know little or nothing about the real history of Luddism. Faith in technology is built into nearly every part of the curriculum as a kind of blind acceptance of the notion of progress. When pressed, however, true believers describe progress to mean not human, political, or cultural improvement but a mindless, uncontrollable technological juggernaut, erasing ecologies and cultures as it moves through history. Technological fundamentalism, like all fundamentalisms, deserves to be challenged. Is technological change taking us where we want to go? What effect does it have on our imagination and particularly on our social, political, and moral imagination? What net effect does it have on our ecological prospects?

George Orwell (1958) once warned that the "logical end" of technological progress "is to reduce the human being to something resembling a brain in a bottle" (p. 201). Behold, 50 years later, there are now those who propose to develop the necessary technology to "download" the contents of the brain into a machine/body (Moravic, 1988). Orwell's nightmare is coming true and in no small part because of research conducted in our most prestigious universities. Such research stands in sharp contrast to our real needs. We need decent communities, good work to do, loving relationships, stable families, the knowledge necessary to restore what we have damaged, and ways to transcend our inherent self-centeredness. Our needs, in short, are those of the spirit; yet, our imagination and creativity are overwhelmingly aimed at things that as often as not degrade spirit and nature.

❖ Conclusion ❖

Ecological education, in Leopold's (1966) words, is directed toward changing our "intellectual emphasis, loyalties, affections, and convictions" (p. 246). It requires breaking free of old pedagogical assumptions, of the straitjacket of discipline-centric curriculum, and even of confinement in classrooms and school buildings. Ecological education means changing (a) the substance and process of education contained in curriculum, (b) how educational institutions work, (c) the architecture within which education occurs, and most important, (d) the purposes of learning.

SOURCES

Bok, D. 1990. *Universities and the Future of America*. Durham: Duke University Press.

Gray, J. G. 1984. *Re-Thinking American Education*. Middletown: Wesleyan University Press.

Griffin, D. 1992. *Animal Minds*. Chicago: University of Chicago Press.

Havel, V. 1992. *Summer Meditations*. New York: Knopf.

James, W. 1987. The Ph.D. Octopus. In B. Kuklick, ed., *William James: Writings 1902–1920*. New York: Library of America. (Original work published 1903.)

Leopold, A. 1966. *A Sand County Almanac*. New York: Ballantine. (Original work published 1949.)

Logsdon, G. 1994. *At Nature's Pace*. New York: Pantheon.

Moravic, H. 1988. *Mind Children*. Cambridge: Harvard University Press.

Newman, J. H. 1982. *The Idea of a University*. Notre Dame, IN: Notre Dame University Press.

Orwell, G. 1958. *The Road to Wigan Pier*. New York: Harcourt Brace Jovanovich.

Pelikan, J. 1992. *The Idea of the University: A Reexamination*. New Haven: Yale University Press.

Polanyi, M. 1958. *Personal Knowledge: Towards a Post-Critical Philosophy*. New York: Harper.

Rowe, S. 1990. *Home Place: Essays on Ecology*. Edmonton, Alberta, Canada: NeWest.

Smith, P. 1984. *Dissenting Opinions*. San Francisco: North Point Press.

Smith, P. 1990. *Killing the Spirit*. New York: Viking Press.

Whitehead, A. N. 1967. *The Aims of Education*. New York: Basic Books. (Original work published 1929.)

Wilson, E. O. 1984. *Biophilia*. Cambridge: Harvard University Press.

CHAPTER FOUR

The Business of Education

AMERICANS presently seem not to agree very much. However, they do appear to agree that public schools are failing badly. On one side of the debate are those, mostly professional educators, who believe that the problem stems from inadequate funds to pay for higher teachers' salaries, better curricula, updated laboratories, newer buildings, and well-stocked libraries. Others have arrived at a different view—a variation on the theme that government is the problem, not the solution. They believe that the public cannot solve its problems publicly. At the extreme, they may believe that a public does not exist at all, only consumers. In this view social problems, like the problems of poor education, cannot be solved except through the profit motive, private ownership, and the magic of free enterprise. Having helped in no small way to starve and demoralize the public sector, this theory is no longer implausible. Acolytes of this view propose more business–education partnerships, more private schools, and a lot more technology. Plato or Thomas Jefferson would scarcely have recognized the reasons being given for educational reforms, which mostly aim to make our young scholars a "world-class work force" in order to make our economy more competitive in international markets. It is American brand names that we want on the next generation of land-filled consumer trash and junk, not those of other countries.

To this end, corporate and business interests have set about to remake education. Something called the New American Schools Development Corporation, created at former President Bush's request and reportedly run by executives on loan from American Telephone and Telegraph, General Motors, Xerox, and other corporate enterprises, is attempting to raise some millions of dollars to finance the creation of innovative educational ideas. They are trying to "rebuild the whole system" in order to "reach the performance of a Toyota or Honda" (Stout, 1992). To do so, the corporation proposes much greater involvement by "the private sector," the health of which, one should note, often requires sizable public subsidies and tax privileges. *Forbes* magazine, for example, devoted its October 14, 1991, issue to the theme "Educating America: An Entrepreneurial Approach." Not unexpectedly, it was longer on advertisements, electronic pedagogy, and hyperventilation than on substance. But having substance or not, business is entering education, and we should take note.

Of course, monied interests have always bought their way into education in the United States and elsewhere. As a result our students often know a great deal more about the growth economy than they do about the economy of nature. The large number of recently endowed business professorships on college campuses suggests that those with money remain dutiful to their interests and to the existence of opportunities through which a little investment can go a long way. By sharing profits, some corporations have been able to annex entire academic departments for research and development. In addition, university administrators, operating expensive institutions and strapped for cash, have been no less alert to these opportunities.

Until recent years, however, business and commercial interests mostly stayed out of public schools. This is no longer the case. Several years ago Whittle Communications Corporation offered public schools free access to Channel One in return for the right to run two minutes of commercial advertising during the school day. Channel One now reaches an estimated 8 million students. Whittle Communications intends to create 1,000 for-profit public schools (the Edison Project), and has hired Benno C. Schmidt, Jr., former President of Yale University, to head the effort.

The Edison Project is the creation of Chris Whittle, founder of Whittle Communications Corporation in Knoxville, Tennessee. He and his partners, among them, Time-Warner, Inc., intend to create a nationwide system of for-profit schools funded by tuition pegged at roughly the national average expenditure per student ($5,500). To make a profit on anticipated annual revenues of $10 billion, Whittle intends to reduce the educational bureaucracy, use a lot of educational technology and fewer teachers, employ students to clean up, ask for volunteers, and rely on economies of scale (*The New York Times*, 1992). The creation of a nationwide voucher system that allows parents to choose public or private schools will help Edison schools greatly. Coincidentally, such a system was supported by former President Bush and his Secretary of Education, Lamar Alexander, a good friend of Mr. Whittle's. Mr. Alexander's stock in Whittle Communications was for a time prudently transferred to Mrs. Alexander, who has no publicly acknowledged connection to Mr. Whittle (Friedman, 1992).

It would be churlish and wrongheaded for me to insist that public education has been a success everywhere and in every way. It has been a mixed bag and for a mixture of reasons. Some public schools have performed quite well, some abhorrently, and most fall somewhere in between. But before tossing babies out with the bathwater, it would be wise for those of us concerned about conserving the earth's biota to ask a few questions. Here is my list.

First, just what do reformers think ails public schools? Benno Schmidt, for one, cites high dropout rates and stagnant or falling Scholastic Aptitude Test scores. "The nation's investment in educational improvement." Schmidt (1992) wrote, "has produced very little return." He attributed these failings to "the system itself" by which he means the following:

> Schools operate around an agrarian calendar that keeps them closed most of the time. They are organized into grades and classrooms that have looked much the same for generations. They are built on a static, factory-like model of one teacher dealing with 20 to 30 children. . . . They use paper, pencils, blackboards and books untouched by modern technology. They reach children only after their earliest formative years. They involve parents only sporadically and casually. They vary little from place to place. . . . They are governed by unwieldy bureaucracies and divided along educationally irrelevant political lines. They are insulated from competition, freedom of choice and innovation. Above all, they are remarkably resistant to change.

This too is a mixed bag, some true, some not so true. Without defending public schools too much, I have a different view. Commercial television, the breakdown of families, the neglect of urban problems, and the culture of violence have made the task of nurturing young minds and hearts far more difficult than it once was. The children we send off to school each morning have been caught in too many domestic quarrels, they have seen too much televised trash, and they have marinated overly long in the me-first consumer culture. It is easy to blame the schools, but this is not primarily a failure of schools. Rather it is a failure of off-duty parents, communities that are no longer communal, presidents and public officials who are too busy to care, and even business executives who are more intent on their profits than on building good communities. How do we create good schools without first creating a good society that values the life of the mind and lives lived with heroism and high purpose?

Second, what qualities of mind and heart do Whittle, Schmidt, and other corporate reformers propose to cultivate? Schmidt (1992) talks about beginning education at "six months [of age] or six months before . . . [babies are] born" a novel, perhaps premature, idea. Is his intent to create "little virtuosos of calculation and competition," in Douglas Sloan's (1993, pp. 117–138) words, or persons with greater depth? What about the sense of wonder? What about the cultivation of more noble qualities, such as honesty and a good heart?

Third, what do the creators of for-profit schools intend to teach our children about biological diversity? About how the earth works? About our ethical responsibilities to other life forms and to future generations? Or about justice? What will they teach about civic responsibility and community? Will they teach critical thinking when it applies to the behavior of the same corporations that are stockholders in Edison schools? Will corporate investors censor what they find distasteful, inconvenient, or that which undermines their profits?

Fourth, how do they propose to teach? Schmidt and Whittle say that they intend to replace a great many teachers with a great deal of technology. But what is known about the relative effects on young people of machines as opposed to caring, well-prepared, and devoted teachers? How will "interactive technology" and computers affect the content of the curriculum? What will be left out because it does not fit a machine-driven pedagogy? And will teachers become just machine tenders? If so, who will children look to as role models?

Fifth, what happens when there is a crunch between education and profits? That crunch will occur, as it does for all businesses. Will investors value the lives and minds of children more than their bottom line? These are not on the same line of the social accounting ledger sheet. If investors do not value education before profits, will we still have enough good public schools to pick up the pieces?

Sixth, how do for-profit schools intend to prepare children to live that portion of their lives that is necessarily public and communal? Can for-profit schools create citizens? Will students participate in school governance? Will for-profit schools prepare children to live and cooperate in a complex world in which many things must be settled on the basis of principle not profit? Will such institutions teach the difference?

Seventh, what will for-profit institutions teach about human nature and about our human potentials? Can for-profit schools teach more than self-interest cleverly pursued? What will happen to the moral imagination of the young entrusted to Edison schools?

Eighth, what will for-profit schools mean for diverse communities in diverse localities with differing traditions? Whittle and Schmidt talk about the economies of scale derived from a nationwide system operating a common curriculum, "completely tied together technologically." Put this way, Edison schools sound a bit like Soviet collective farms with power concentrated in the hands of a few administrators and investors. What will such schools do to local traditions and local knowledge? Will they foster a sense of place or rootlessness?

Finally, what kind of world do the creators of for-profit schools intend to create? Is it a world that will conserve biological and cultural diversity? Do they intend to create a civilization that fits within the biosphere? Do they intend to do so with justice and fairness? If so, how? And how might for-profit schools regard the agenda of the twenty-first century, which requires that we stabilize climate, become much more efficient in the use of fossil fuels, make a rapid transition to solar-based technologies, stabilize and then reduce population, reverse the loss of forests, protect biodiversity, conserve soil, rebuild rural areas, clean up toxic messes, and sharply reduce poverty?

Those proposing for-profit schools have not said much that would give us reason to believe that they have thought about such things. They seem to think of education as a technical problem solvable by more technology. They speak the language of business talking of "investments," "risk taking," "entrepreneurship," "competition," and "performance," as if education were a matter of profit and loss. When losses get too high, companies cut their losses, but communities and societies cannot. Those intending to create for-profit schools are part of the new conventional wisdom that holds that privatization is the answer to public problems. In some circumstances, it may be, but not in education. The answer to poor schools is to create better communities that take their children and their long-term prospects seriously regardless of the cost.

SOURCES

Friedman, J. 1992, February 17. Big Business Goes to School. *The Nation*, pp. 188–192.

The New York Times. 1992, May 26, p. A12.

Schmidt, B. C. 1992, June 5. Educational Innovation for Profit. *The Wall Street Journal.*

Sloan, D. 1993. A Postmodern Vision of Education for a Living Planet. In D. R. Griffen and R. Falk, eds., *Postmodern Politics for a Planet in Crisis.* Albany: State University of New York Press.

Stout, H. 1992, February 14. Teams Vie to Redesign U.S. Education. *The Wall Street Journal.*

PART TWO

FIRST PRINCIPLES

IF TINKERING reforms are not an adequate response to our plight—and they are not—we must rethink our initial assumptions about learning and the goals of education. The essays in Part two were written in the belief that we have lost sight of fundamental things that are essential to this effort. As the curriculum has become more extensive, complex, and technologically sophisticated, important things are being lost. I am not referring to what some describe as "basics," such as the ability to read, write, and count, although these too, I suppose, are at risk. I am referring to the relationship between an increasingly specialized educational process and our ability to ask large questions having to do with the human condition. These essays, then, have to do with love, intelligence, wisdom, virtue, responsibility, value, and good sense.

CHAPTER FIVE

Love

We cannot win this battle to save species and environments without forging an emotional bond between ourselves and nature as well—for we will not fight to save what we do not love.
— S. J. GOULD

STEPHEN Jay Gould's is one view of the issue, and in most academic disciplines, a decidedly minority view. Mainstream scholars who trouble themselves to think about disappearing species and shattered environments appear to believe that cold rationality, fearless objectivity, and a bit of technology will get the job done. If that were the whole of it, however, the job would have been done decades ago. Except as pejoratives, words such as *emotional bonds*, *fight*, and *love* are not typical of polite discourse in the sciences or social sciences. To the contrary, excessive emotion about the object of one's study is in some institutions a sufficient reason to banish miscreants to the black hole of committee duty or administration on the grounds that good science and emotion of any sort are incompatible, a kind of presbyterian view of science.

Gould's view raises a number of questions. For example, how is it possible to reconcile the procedural requirements of science with those of emotional bonding and love for the creation, a process that Gould believes to be necessary to save species and environments? Do these work at cross purposes? Do they entail different curricula and different kinds of education from that now offered almost everywhere? For survival purposes, must curiosity be tempered by affection? Might the same be equally true for physics and economics? Assuming that we could define *love* satisfactorily, would it set limits on knowledge or on the way in which knowledge is acquired? In any conflict between the requirements of love and those of knowledge, which should have priority? Inherent in Gould's view is the paradox that we must learn to act selflessly, as sages and religious prophets have said all along, but for reasons that are increasingly indistinguishable from our self-interest in survival.

These are important and difficult issues on which reasonable people can disagree, but the stakes are high enough to warrant serious and sustained discussion about love in relation to education and knowledge. But hardly a whisper of that debate is evident. In a cursory survey of indexes to biology texts, for example, I found no mention of the word *love*. Neither did I find any entry under *emotional bonding* or *fight* for that matter. The same was true of other textbooks in physics, chemistry, political science, and economics. Why is it so hard to talk about love, the most powerful of human emotions, in relation to science, the most powerful and far-reaching of human activities? And why is this so for textbooks written to introduce the young to the disciplined study of life and life processes? That place of introduction would appear to be a good point at which to say a few words about love, awe, and mystery and perhaps a caution or two about the responsibilities that go with knowledge. This might even be a good place to discuss emotions in relation to intellect and how best to join the two, because they are joined in one way or another. It is as if there were a conspiracy of silence about what drives the effort to acquire knowledge. Perhaps it is only embarrassment about what does or does not move us personally.

There are other reasons as well why we have failed to think much about the issue of love in relation to science and education. Undoubtedly, the loudest objection to any such discussion will be made by the more rigorous than thou, the academic equivalent of the fundamentalist, who will argue that science works inversely to passion, their own passion for purity notwithstanding. Yet their science, in psychologist Abraham Maslow's (1966) words, can be

> used as a tool in the service of a distorted, narrowed, humorless, de-eroticized, de-emotionalized, desacralized and desanctified Weltanschauung. This desacralization can be used as a defense against being flooded by emotion, especially the emotions of humility, reverence, mystery, wonder, and awe. (p. 139)

Fundamentalists have mistaken the relation between passion, emotion, and good science. These are not antithetical, but complexly interdependent. Science, at its best, is driven by passion and emotion. We have emotions for the same reason we have arms and legs: They have proved to be useful over evolutionary time. The point in either case is not to cut off various appendages and qualities, but rather to learn to coordinate and discipline them to good use.

The problem with scientific fundamentalism is that it is not scientific enough. It is rather a narrow-gauge view of things that is ironically unskeptical, which is to say, unscientific, about science itself and the larger social, political, economic, and ecological conditions that permit science to flourish in the first place.

Second, for all of our information and communications prowess, we talk too little about our motives and feelings in relation to our occupations and professions. I recall, for example, a conversation in which a group of distinguished ecologists and environmentalists was asked to describe the sources of their beliefs. In trying to describe their deepest emotions as if they were the result of carefully considered career plans, these otherwise eloquent people descended into a pit of muddled incoherence. But as the conversation continued, deeply moving stories about experiences of the most personal kind began to emerge. Most of us have had similar experiences. But we tend to talk about "career decisions" as if our lives were rationally calculated and not the result of likes, fascinations, imaginative happenings, associations, inspirations, and sensory experiences stitched into our childhood or early adult memories. I believe that most of us do what we do as environmentalists and profess what we do as professors because of an early, deep, and vivid resonance between the natural world and ourselves. We need to be more candid with ourselves and our students who have chosen to study biology or the human place in the environment because of a similar resonance.

Third, I think the power of denial in a time of cataclysmic changes undermines our willingness to talk about important things. There is a scene, for example, in the movie *The Day After* in which a woman, knowing that an H-bomb is about to hit, scurries about to tidy things up. A good bit of what goes on in the modern university likewise seems to me like a kind of tidying up before all hell breaks loose. Of course, it has already broken loose, and more is on the way. The twentieth century is the age of world wars, atomic bombs, gulags, totalitarianism, death squads, and ethnic cleansing. Looking ahead, we see the threats of biotic impoverishment, changing climate, and overpopulation. In the light of such prospects, it is understandable that many find it easier and safer to tidy things up rather than roll up their sleeves to turn these trends around.

Fourth, love is difficult to talk about because we do not know much about how it is built into our deepest emotions and exhibited in our various behaviors. We have good reason, however, to believe that what Edward O. Wilson (1992) calls "biophilia," or "the connections that humans subconsciously seek with the rest of life," is innate (p. 350). It would be surprising indeed if several million years of evolution had resulted in no such affinity. But even if we could find no trace of subconscious biophilia, our concern for survival would cause us to invent it for the same reason Gould (1991) described above: We are not likely to fight to save what we do not love. This means that biophilia must become a conscious part of what we do and how we think, including how we do science and how we educate people to think in all fields. What does this mean for teachers and scholars?

For one thing, it means that biophilia, conscious and subconscious, deserves to be a legitimate subject of conversation and inquiry (Kellert and Wilson, 1993). We need to become students of biophilia in order to understand more fully how it comes to be, how it prospers, and what it requires of us. For another, it requires a greater consciousness about how language, models, theories, and curricula can sometimes alienate us from our subject matter. Words that render nature into abstractions of board feet, barrels, sustainable yields, and resources drive out such feelings and the affinities we have at a deeper level. We need better tools, models, and theories, calibrated to our innate loyalties—ones that create less dissonance between what we do for a living, how we think, and what we feel as creatures who are the product of several million years of evolution.

Finally, and most difficult, from other realms we know that love sets limits to what we do and how we do it. It is, as Erich Fromm (1956) once wrote, an art requiring "discipline, concentration, and patience" (p. 100). What does the art of love have to do with the discipline of science? On one side of the question, love is not a substitute for careful thought. On the other side, when the mind becomes, in Abraham Heschel's (1951) words, "a mercenary of our will to power . . . trained to assail in order to plunder rather than to commune in order to love" (p. 38), ruin is the logical result. In either case it is evident that personal motives matter, and different motives lead to very different kinds of knowledge and very different ecological results.

At a recent meeting of conservation biologists, some of the participants wondered out loud why so few of their colleagues had joined the effort to conserve biological diversity. No good answer was given. On reflection, however, I think the reason lies in the difficulty we have in joining professional science with our love of life and those things that probably attracted most biologists to study science in the first place and continue to attract our students.

SOURCES

Fromm, E. 1956. *The Art of Loving*. New York: Harper.

Gould, S. J. 1991, September. Enchanted Evening. *Natural History*, p. 14.

Heschel, A. J. 1951. *Man Is Not Alone: A Philosophy of Religion*. New York: Farrar, Straus & Giroux.

Kellert, S., and Wilson, E., eds. 1993. *The Biophilia Hypothesis*. Washington, DC: Island Press.

Maslow, A. 1966. *The Psychology of Science*. Chicago: Regnery.

Wilson, E. O. 1992. *The Diversity of Life*. Cambridge: Harvard University Press.

CHAPTER SIX

Some Thoughts on Intelligence

Ours is about the most ignorant age that can be imagined.
— ERWIN CHARGAFF

NO OTHER society, to my knowledge, has made such a fetish of intelligence as has modern America. Indeed we have what philosopher Mary Midgley calls a veritable "cult of intelligence" administered by tribes of experts whose function is to measure it, raise it, write books about it, and make those without it feel bad. But exactly what is intelligence? More to the point, what is intelligence as measured against the standards of biological diversity and human longevity on earth? And what might such answers say about how we go about the work of conservation education?

I have no credentials to raise such questions. I could not distinguish the Minnesota Multiphasic Personality Inventory from multiple phases in Minnesota. Nonetheless, I believe that, in the main, the evidence now indicates that we do not know very much about intelligence and that from the perspective of the earth, much of what we presume to know may be wrong, which is to say that it is not intelligent enough. I am also inclined to agree that:

> The modern stereotype of an intelligent person is probably wrong. The prototypical modern intelligence seems to be that of the Quiz Kid—a human shape barely discernable in [a] fluff of facts. (Berry, 1983, p. 77)

What we call intelligence and what we test for and reward in schools and colleges is something else, more akin to cleverness. Intelligence, as I

'erstand it, has to do with the long run and is mostly integrative,
as cleverness is mostly preoccupied with the short run and tends to
t things. This distinction has serious consequences for our will-
nd ability to conserve biological diversity.

Although I do not think it is possible to give an adequate definition
of intelligence, I believe it is possible to describe certain characteristics of
it. First, people acting or thinking with intelligence are good at separating
cause from effect. Geographer I. G. Simmons, for example, tells the story
of an eighteenth-century protopsychiatrist who developed an infallible
method of distinguishing the sane from the insane. Those to be diagnosed
he locked in a room with water taps on one side and a supply of mops
and buckets on the other. He then turned on the taps and watched: Those
he considered mad ran for the mops and buckets; the sane walked over
and turned off the taps (Simmons, 1989, p. 334). I keep a file of "mop
and bucket" proposals by persons well paid and honored for their reputed
intelligence, the contents of which range from the ridiculous through the
absurd to the potentially criminal. The common characteristic of these is
the recurrent inability to ask questions having to do with big causes and
large consequences.

A second and related characteristic of intelligence is the ability to separate "know how" from "know why." Canadian ecologist Stan Rowe
(1990), for example, quoted Enrico Fermi, saying to a skeptic before the
first atomic test, "Don't bother me with your conscientious scruples; after
all, the thing's superb physics!" (p. 129). I also understand that some
believed that the detonation of an atomic bomb might set off a chain reac-
tion that would destroy the entire planet. It was done nonetheless. However superb the physics, the human and ecological consequences of the
bomb have been disastrous and we have not seen the last of it. The bomb
is only one illustration of the kind of obsessiveness that feeds on cleverness
without internal or external restraint and, in this instance, with governmental sanction. We are capable of doing many more clever things than
intelligence would have us do. But obsession to do whatever is possible
regardless of whether it is desirable is no longer unusual in the modern
university, which has become, in Rowe's (1990) words, a "know how"
rather than a "know why" kind of institution. Its stock in trade is "ignorant knowledge," by which Rowe means "knowing in fragments, knowing without direction, knowing without commitment" (p. 129). Its
graduates "score high on means tests but low on ends tests." And the
means are derived from paradigms that are "dead wrong . . . life-denying,

fostering sickness instead of health." Intelligent people would reverse the order, first asking "why?" and "for what reason?" Real intelligence often works slowly and is close to or even synonymous with what we call wisdom. Intelligent people, according to Mary Midgley (1990),

> excel in a different range of faculties. . . . They possess strong imaginative sensibility—the power to envisage possible goods that the world does not yet have and to see what is wrong with the world as it is. They are good at priorities, at comparing various goods, at asking what matters most. They have a sense of proportion, and a nose for the right directions. (p. 41)

People possessed of such faculties, Midgley calls "wise or sensible rather than clever or smart." And they know when enough is enough.

Wendell Berry (1983) suggests a third characteristic of intelligence: the "good order or harmoniousness of his or her surroundings." By this standard "any statistical justification of ugliness or violence is a revelation of stupidity" (p. 77). This means that the consequences of one's actions are a measure of intelligence and the plea of ignorance is no good defense. Because some consequences cannot be predicted, the exercise of intelligence requires forbearance and a sense of limits. In other words it does not presume to act beyond a certain scale on which effects can be determined and unpredictable consequences would not be catastrophic. In Berry's words, intelligent people (and civilizations) do not assume "that we can first set demons at large, and then, somehow become smart enough to control them" (p. 65). If there is such a thing as a societal IQ, what we call "developed" societies would be judged retarded by Berry's standard. Overflowing landfills, befouled skies, eroded soils, polluted rivers, acidic rain, and radioactive wastes suggest ample attainments for admission into some intergalactic school for learning-disabled species.

A fourth characteristic of intelligent action and thought is that it does not violate the bounds of morality. In other words, it does not, in the name of some alleged higher good, demand the violation of life, community, or decency. Intelligent behavior is consonant with moderation, loyalty, justice, compassion, and truthfulness, not for ethereal theological reasons but because these are fundamental to living well. Morality is long-term practicality that recognizes our limits, fallibility, and ignorance. On the other hand, the cultivation of vice leads, as E. F. Schumacher (1973) once wrote, to "a collapse of intelligence." Driven by vice a person "loses the power of seeing things as they really are . . . in their roundness and wholeness" (p. 29). Intellect driven by vice cannot lead to intelligent action or thought. Said differently, real intelligence depends upon character as much as it does on mental horsepower. The corruption of character, as Emerson wrote in his essay "Nature" (1836), leads in turn to "the corruption of language . . . new imagery ceases to be created, and old words are perverted to stand for things which are not. . . . In due time the fraud is manifest, and words lose all power to stimulate the understanding or the affections" (pp. 32–33).

From these characteristics, I conclude that it is possible for a person to be clever without being very intelligent, or as Walker Percy put it to "get all A's and flunk life." Furthermore, whole civilizations can be simultaneously clever and stupid by which I mean that they might be able to perform amazing technological feats while being unable to solve their most basic public problems. Perhaps these go together. As Exhibit A, consider our phenomenal and growing computer capabilities side by side with our decaying inner cities, insensate violence, various addictions, rising public debt, and the destruction of nature all around us. Can it be that we are in fact becoming both more clever and less intelligent? If so, why is this?

I am tempted to round up all of the usual members of the rogue's gallery from Descartes ("I think therefore I am"), through all of the peddlers of instrumental rationality, artificial intelligence, and unfettered curiosity, all of whom are eminently blameworthy. But they are only partial manifestations of a deeper cause having to do with the very origins of intelligence.

Could it be that the "integrity, stability, and beauty" of nature is the wellspring of human intelligence? Could it be that the conquest of nature, however clever, is in fact a war against the source of mind? Could it be that the systematic homogenization of nature inherent in contemporary technology and economics is undermining human intelligence? If so, biological diversity is important to us not only as a source of wonder drugs and miracle fruits, but as the source of what made us human. We have good reason to believe that human intelligence could not have evolved in a lunar landscape, devoid of biological diversity. We also have good reason to believe that the sense of awe toward the creation had a great deal to do with the origin of language and why protohumans *wanted* to talk, sing, and write poetry in the first place. Elemental things like flowing water, wind, trees, clouds, rain, mist, mountains, landscape, animal behavior, changing seasons, the night sky, and the mysteries of the life cycle gave birth to thought and language. They continue to do so, but perhaps less exuberantly than they once did. For this reason I think it not possible to unravel the creation without undermining human intelligence as well. The issue is not so much about what biodiversity can do for us as resources as it is about the survival of human intelligence cut off from its source.

Cleverness would have us advance a narrowly defined, short-term, and anemic self-interest at all costs and at all risks. But cleverness, pure intellect, is just not intelligent enough. Its final destination is madness. Intelligence would lead us, on the contrary, to protect biological diversity, but for reasons that go beyond the calculation of self-interest. The surest sign of the maturity of intelligence is the evolution of biocentric wisdom, by which I mean the capacity to nurture and shelter life—a fitting standard for a species calling itself *Homo sapiens*.

What can educators do to foster real intelligence? One view is that we should not try, because the best we can do is to help students avoid being stupid (Postman, 1988, p. 87). I think we should prevent stupidity where possible, but I also think we can do more. First, we can question the standard model of pre-ecological intelligence and encourage students to think the matter out for themselves, including the matter of collective intelligence. Second, we can reward intelligence in all sorts of ways without necessarily penalizing cleverness. Third, we can develop the kind of first-hand knowledge of nature from which real intelligence grows. This means breaking down walls made by clocks, bells, rules, academic requirements, and a tired indoor pedagogy. I am proposing a jail break that would put learners of all ages outdoors more often. Fourth, we can liberalize the liberal arts to include ecological competence in areas of restoration ecology, agriculture, forestry, ecological engineering, landscape design, and solar technology. Fifth, we can suspend the implicit belief that a PhD is a sign of intelligence and draw those who have demonstrated a high degree of applied ecological intelligence, courage, and creativity (farmers, foresters, naturalists, ranchers, restoration ecologists, urban ecologists, landscape planners, citizen activists) into education as mentors and role models. Finally, we can attempt to teach the things that one might imagine the earth would teach us: silence, humility, holiness, connectedness, courtesy, beauty, celebration, giving, restoration, obligation, and wildness.

SOURCES

Berry, W. 1983. *Standing by Words*. San Francisco: North Point Press.

Emerson, R. W. 1972. *Selections from Ralph Waldo Emerson*. S. Whicher, ed. Boston: Houghton Mifflin. (Original work published 1839.)

Midgley, M. 1990. Why Smartness is Not Enough. In M. Clark and S. Wawrytko, eds., *Rethinking the Curriculum*. Westport: Greenwood Press.

Postman, N. 1988. *Conscientious Objections*. New York: Knopf.

Rowe, S. 1990. *Home Place: Essays on Ecology*. Edmundton, Alberta, Canada: NeWest.

Schumacher, E. F. 1973. *Small is Beautiful: Economics as if People Mattered*. New York: Harper Torchbooks.

Simmons, I. G., 1989. *Changing the Face of the Earth*. Cambridge, England: Blackwell.

CHAPTER SEVEN

Reflections on Water and Oil

T HE MEANING of water might best be approached in comparison with that other liquid to which we in the twentieth century are beholden: oil. Water as rain, ice, lakes, rivers, and seas has shaped our landscape. But oil has shaped the modern mindscape, with its fascination and addiction to speed and accumulation. The modern world is in some ways a dialogue between oil and water. Water makes life possible, while oil is toxic to most life. Water in its pure state is clear; oil is dark. Water dissolves; oil congeals. Water has inspired great poetry and literature. Our language is full of allusions to springs, depths, currents, rivers, seas, rain, mist, dew, and snowfall. To a great extent our language is about water and people in relation to water. We think of time flowing like a river. We cry oceans of tears. We ponder the wellsprings of thought. Oil, on the contrary, has had no such effect on our language. To my knowledge, it has given rise to no poetry, hymns, or great literature, and probably to no flights of imagination other than those of pecuniary accumulation.

Our relation to water is fundamentally somatic, which is to say it is experienced bodily. The brain literally floats on a cushion of water. The body consists mostly of water. We play in water, fish in it, bathe in it, and drink it. Some of us were baptized in it. We like the feel of salt spray in our faces and the smell of rain that ends a dry summer heat wave. The sound of mountain water heals what hurts. We are mostly water and have an affinity for it that transcends our ability to describe it in mere words.

Oil and water have had contrary effects on our minds. Water, I think, lies at the origin of language. It is certainly a large part of the beauty of language. Water has also given rise to some of our most elegant technologies: water clocks, sailing ships, and waterwheels. The wise use of water is quite possibly the truest indicator of human intelligence, measurable by what we are smart enough to keep out of it, including oil, soil, toxics, and old tires. The most intelligent thing we could have done with oil was to have left it in the ground or to have used it very slowly over many centuries. Oil came to western civilization as a great temptation to binge, devil take the hindmost. Our resistance had already been lowered by the intellectual viruses introduced by the likes of Galileo, Bacon, and Descartes. We were in no condition to fend off those introduced by John D. Rockefeller, Henry Ford, and Alfred P. Sloan that promised speed, mobility, sexual adventure, and personal identity. Oil has undermined intelligence in at least six ways.

First, oil eroded our ability to think intelligently about community and the possibility of cooperation. Its nature is what game theorists call zero-sum: You have it or someone else does; you burn it or they do. Its possession set those who had it against those who did not: states against states; regions against regions; nations against nations; and the interests of one generation against those of generations to follow. Cheap oil and the automobile pitted community against community, suburban commuters against city neighborhoods. Money made from oil and oil-based technologies corrupted our politics, while our growing dependency corrupted our sense of proportion and scale. To guarantee our access to Middle Eastern oil we have declared our willingness to initiate Armageddon. We are now spending billions in fulfillment of this pledge even though a fraction of this annual bill would eliminate the need for oil imports altogether. The characteristics of oil and the way we have used it and have grown overly dependent on it have helped shape a mind-set that cannot rise above competition.

Second, oil has undermined our land intelligence by increasing the speed with which we move on it or fly over it. We no longer experience the landscape as a vital reality. Compare a trip by interstate highway from Pennsylvania to Florida with that taken by William Bartram in the eighteenth century. Where Bartram saw wonders and had the time to observe them carefully and be instructed and moved by them, modern travelers experience only a succession of homogenized images and sounds moving through an engineered landscape all tailored to the requirements of speed and convenience. As a result, our contact with land is increasingly abstract, measured as lapsed time and experienced as the dull exhaustion that accompanies jet lag or close confinement.

Third, oil has made us dumber by making the world more complicated but less complex. An Iowa cornfield is a complicated human contrivance resulting from imported oil, supertankers, pipelines, commodity markets, banks and interest rates, federal agencies, futures markets, machinery, spare parts supply systems, and agribusiness companies that sell seeds, fertilizers, herbicides, and pesticides. In contrast, the forest or prairie that once existed in that place was complex, a highly resilient system consisting of a diversity of life forms, ecological relationships, and energy flows. Complicatedness is the result of high energy use that destroys genetic and cultural information. With complicatedness has come specialization of knowledge and the "expert." Exit the generalist and the renaissance person. The result is a society and economy that no one comprehends, indeed, one that is beyond human comprehension. Complicatedness gives rise to unending novelty, surprise, and unforeseen consequences. As the possibility of foresight declines, the idea of responsibility also declines. People cannot be held accountable for the effects of actions that cannot be foreseen. Moreover, a high-energy society undermines our sense of meaning and our belief that our own lives can have meaning. It leads us to despair and to disparage the very possibility of intelligence.

Fourth, cheap oil and the automobile are responsible, in large measure, for the urban sprawl that has conditioned us to think that ugliness and disorder are normal or at least economically necessary. Where fossil energy was cheap and abundant the idea of a land ethic based on the "integrity, stability, and beauty of the biotic community" has never taken firm hold. This is not just a problem of ethics; it is a deeper problem that has to do with how poorly we think about economics. Sprawling megalopolitan areas are not only an aesthetic affront; they are sure signs of an unsustainable economy dominated by absentee corporations that vandalize distant places for "resources" and other places to discard wastes. A mind conditioned to think of ecological, aesthetic, and social disorder as normal, which is to say a mind in which the categories of harmony and beauty have atrophied, is to that extent impoverished. It is rather like being able to see only half of the color spectrum. On the other hand, intelligence, I think, grows as the mind is drawn to the possibilities of creating order, harmony, beauty, stability, and permanence.

Fifth, oil has undermined intelligence by devaluing handwork and craftsmanship. To a great extent the history of high-energy civilization can be described by the shift in the ratio between labor and energy. Economic development is the process of substituting energy for labor, moving people from farms into cities and from craft professions into factories and eventually into "the service sector." This is not simply a matter of economic efficiency as some argue; it is a problem of human intelligence. Thinking, doing, and making exist in a complex symbiotic relationship. The price we pay for the convenience and affluence of a service economy may well be paid in the coin of intelligence. The drift of high-energy civilization is to make the world steadily less amenable to the kind of thought that results from the friction of an alert mind's grappling with real materials toward the goal of work well done. To the modern mind, with its ghettos of costs and benefits, expertness, efficiency, built-in obsolescence, and celebration of technology that replaces manual skill, any alternative sounds hopelessly naive. However, we may find reason to reconsider, on the grounds of a larger efficiency and higher rationality, the reality that we are in fact "homo faber" whose identity is defined by the close interplay of thought and making.

Finally, oil has undermined intelligence because it requires technologies that we are smart enough to build but not smart enough to use safely. This is the gap between knowing how to do something and knowing what one should do. Cheap oil has divided our capabilities from our sense of obligation, care, and long-term responsibility. Oil used at the rate of millions of barrels each day cannot be used responsibly. The *Exxon Valdez* oil spill in Prince William Sound, and the dozens of other large oil spills like it, are not accidents but the logical result of a system that operates on a scale that can only produce catastrophes. Our mistake is compounded by the belief that the catastrophe occurred only because oil was spilled. It would have been an equal, if more diffuse catastrophe, had the *Exxon Valdez* made it safely to port and its cargo burned in car engines, proceeding thence into the atmosphere where its contents would have contributed to air pollution and global warming. Oil has reduced our intelligence by dividing us between what we take to be realistic imperatives of economy and the commands of ethical stewardship. As a result we have become far less adept at thinking and acting ethically and far more adept at rationalizing and denying.

If oil has made us dumber, might water make us smarter about more things over a longer term? I think so. To this end, I suggest several things beginning with an examination of contemporary curriculum to identify those parts that are based on the assumption of the permanence and blessedness of cheap energy. How much of the curriculum would stand if this assumption were removed? Education has generally prepared the young to live in a high-energy world. We have shaped whole disciplines around such assumptions without stopping to inquire about their validity or their larger effects. The belief in the permanence and felicity of high-energy civilization is found at the heart of most of contemporary economics, with its practice of discounting, development theory, marketing, business, political science, and sociology. The natural sciences have been largely directed toward manipulation of the natural world without any comparable effort to study impacts of doing so or alternative kinds of knowledge that work with natural systems. Behind a great deal of this is the belief that we can make an end run around nature and get away with it.

Second, water should be a part of every school curriculum. I would include, for example, Karl Wittfogel's (1956) study of the relationship between water management and despotic government, Donald Worster's (1985) study of the politics of water in the American West, and Charles Bowden's (1985) study of the relationship between water and the Papago people of Arizona. Water as part of our mythology, history, politics, culture, and society should be woven throughout curriculum, K through PhD.

Third, water should be the keystone in a new science of ecological design. John Todd's (1991) *Living Machines* is a working example of ecological design. Education in ecological design would have to be transdisciplinary, aiming to integrate a broad range of disciplines and design principles of resilience, flexibility, appropriate scale, and durability. Todd's work, as an example, is instructive in part because he has combined good engineering with ecology and vision.

Fourth, water and water purification should be built into the architecture and the landscape of educational institutions. The very institutions that purport to induct the young into responsible adulthood often behave like vandals. This need not be. Institutional waste streams offer a good place to begin to teach applied (as opposed to theoretical) responsibility. Solar aquatic waste systems and similar approaches offer a way to teach the techniques of waste water purification, biology, and closed loop design. There are many reasons to regard resource and waste flows as a useful part of the curriculum, not merely a nuisance.

Finally, I propose that restoration be made a part of the educational agenda. Every public school, college, and university is within easy reach of streams, rivers, and lakes that are in need of restoration. The act of restoration is an opportunity to move education beyond the classroom and laboratory to the outdoors, from theory to application and from indifference to healing. My proposal is for institutions to adopt streams or entire watersheds and make their full health an educational objective as important as, say, capital funds campaigns to build new administration buildings or athletic facilities.

What is the meaning of water? One might as well ask, "What does it mean to be human?" The answer may be found in our relation to water, the mother of life. When the waters again run clear and their life is restored we might see ourselves reflected whole.

SOURCES

Bowden, C. 1985. *Killing the Hidden Waters*. Austin: University of Texas Press.
Todd, J. 1991. *Living Machines*. Unpublished manuscript.
Wittfogel, K. 1956. *Oriental Despotism*. New Haven: Yale University Press.
Worster, D. 1985. *Rivers of Empire*. New York: Pantheon.

CHAPTER EIGHT

Virtue

A T A recent gathering of Washington, DC, environmentalists, the prevailing wisdom held that the public could not be led to a survivable future by moral arguments but only by arguments that appeal to short-term economic self-interest. This is a widely held opinion and one that raises a serious question for conservation educators. Is conservation primarily a technical subject—how to achieve our self-interest with the lowest environmental impacts—with only minor moral implications, or is it fundamentally about morality—what we should want—with minor technical details? If the former, then having equipped our students with a thorough grasp of the pertinent scientific disciplines, the technological basis of efficient resource use, and a bit of economics, we may regard our duties as educators adequately discharged. If the latter, we must do all of the above while enabling students to think clearly about what was once called without apology "virtue" and enable them to live accordingly. The difference between the two is partly that between reform and "perestroika." It is a difference in whether one thinks that with the right technologies and prices we will make a smooth transition to the condition of sustainability or whether we will make the transition, if we make it, by the margin of a gnat's eyebrow with the four horsemen in hot pursuit. On grounds of prudence and my reading of the available evidence, I am persuaded of the latter and hence of the need to think seriously about the relationship between sustainability and the human qualities subsumed in the word *virtue*.

But what is virtue? Philosopher Alasdair MacIntyre believes that the modern world suffers from moral amnesia, the vague awareness of a deficiency of virtue that we can no longer describe. To understand virtue he has argued that we must return to its ancient roots, for "the tradition of the virtues is at variance with central features of the modern economic order and more especially its individualism, its acquisitiveness and its elevation of the values of the market to a central social place" (MacIntyre, 1981, p. 237). As it was understood in the ancient world, virtue was founded on the bedrock of community. One's virtue was inseparable from one's life within a community. From this perspective, according to MacIntyre, "the egoist is thus . . . someone who has made a fundamental mistake about where his own good lies" (p. 213). Robert Proctor (1988) has made the same point in a remarkable book, *Education's Great Amnesia*: "The ancients . . . conceptualized and experienced their humanity not as separation, but as participation in the whole order of being" (p. 166). Virtue was regarded, first, as an exercise in participation and fulfillment of the obligations of membership in a community that was embedded in a larger cosmic order.

A second aspect of ancient virtue was, according to Cicero as translated by Grant (1987), "the ability to restrain the passions and to make the appetites amenable to reason" (p. 128). Moderation, as Aristotle defined it (Oswald, trans., 1962) was the mean between extremes of excess and deficiency that could be defined by a person of practical wisdom. Virtue, for Aristotle, is chosen through the exercise of reason: "It is not possible to be good in the strict sense without practical wisdom, nor practically wise without moral virtue" (p. 172). In other words, virtue is the result of choosing intelligently between extremes.

Third, for the Greeks and the Romans, virtue was never separated from politics and from participation in the civic life of the community. For Aristotle the cultivation of virtue was both a goal of politics—"to engender a certain character in the citizens and to make them good and disposed to perform noble actions" (p. 23)—and a prerequisite for civic order, since no good community could be constructed by people without virtue. Modern politics has rejected that tradition, replacing authority based on virtue with scientific management and public relations.

In the ancient world virtue also meant the cultivation of qualities of courage, fortitude, honesty, restraint, charity, chastity, family, personal rectitude, integrity, and reverence. However imperfectly these were realized in practice, they provided the standard by which people judged themselves. The fact that this list sounds archaic to the modern ear is an indication of how far we have gone in the contrary direction. Modern societies are increasingly operated by and for that subsystem known as the economy, the same economy that, as Lewis Mumford once observed, converted the seven deadly sins of pride, envy, anger, sloth, avarice, gluttony, and lust into virtues after a fashion and the seven virtues of faith, charity, hope, prudence, religion, fortitude, and temperance into sins against gross national product. The dependence of the economy on sin is a phenomenon infrequently studied by economists. Sin, a contentious subject, has been replaced with the more socially agreeable doctrine that all things are relative so that anyone's opinions or behavior are as good as those of any other, or at least not much worse. But lacking the qualities of virtue, can we do the difficult things that will be necessary to live within the boundaries of the earth?

I think not, first, because people lacking a sense of community that undergirds the practice of virtue are not likely to care how their actions affect the larger world in any but the most superficial way. Can we expect rational maximizers of self-interest, who discount the future interests of their own children and grandchildren, to be moved by their kinship to bugs and biota? Not likely. Virtue, as Aristotle and Cicero described it, was founded on a kind of moral ecology (albeit one that excluded lots of people), an awareness of mutual dependence. Lacking this sense, people are not likely to care deeply enough to join the constituency for change that must finally think, live, and vote differently. People who regard their welfare narrowly are unlikely to support large-scale social change when it costs something. Hence, without a virtuous public that cares deeply about the protection and enhancement of life, there will be no constituency for hard choices ahead and for the policy changes necessary for sustainability.

Second, sustainability will require a reduction in consumption in wealthy societies and changes in the kinds of things consumed toward products that are durable, recyclable, useful, efficient, and sufficient. This will occur when enough people choose to consume less or when scarcity is imposed by circumstances and enforced by government fiat, as Robert Heilbroner (1974) once predicted. It will not come about by putting bandages on potentially terminal wounds, making plastics that are biodegradable, for instance. If we are not to turn the earth into a toxic dump or bankrupt ourselves by expensively undoing what should not have been done in the first place, moderation must eventually replace self-indulgence. The appetites, as Cicero put it, must be made "amenable to reason," which for us means making them less amenable to advertising and television.

Third, a great deal has been said about the potential for least-cost, end-use analysis that hitches narrow economic rationality to the efficient use of resources with better technology. This is all to the good. However, problems arise when that same economic rationality causes consumers to observe that least cost is not the same as full cost. For example, the fully informed consumer, armed with least-cost reasoning, would certainly choose to buy compact fluorescent lightbulbs that have lower lifetime costs than incandescent bulbs. But the same narrow economic rationality would cause that consumer to refuse to pay higher utility costs to clean up nuclear wastes and decommission reactors used to generate the electricity that is used with greater efficiency. At this point, economic rationality stops and virtue begins. Least-cost reasoning applies to those costs that must be paid now; full cost applies to those costs that can be pushed onto others or deferred to our children. Only people who take their obligations seriously, people of virtue, would willingly pay the full costs of their actions or even demand to do so.

Fourth, it is implausible, as noted in Chapter 6, that we can systematically cultivate pride, gluttony, lust, avarice, sloth, envy, and anger and remain intelligent. The seven deadly sins are sins in large part because they corrode the intellect. Virtue is a product of reason, not impulse, whim, and fantasy. Anything that destroys the capacity for reasoned choice promotes sin and a grosser national product. On a larger scale, does the deliberate cultivation of sin make us a dumber society? Aristotle would have thought so. And as we become dumber, more passive, and less morally adept, do we also become more tolerant of (less capable of recognizing and being outraged by) malfeasance, arrogance, stupidity, and vacuousness by public officials? As officialdom becomes more corrupt, inept, and shortsighted, can its management of the environment become better? Hardly.

SOURCES

Grant, M., translator. 1987. Cicero, *On the Good Life*. London: Penguin Books.

Heilbroner, R. 1974. *An Inquiry into the Human Prospect*. New York: Norton.

MacIntyre, A. 1981. *After Virtue*. Notre Dame, IN: Notre Dame University Press.

Oswald, M., translator. 1962. Aristotle, *Nichomachean Ethics*. Indianapolis: Bobbs-Merrill.

Proctor, R. 1988. *Education's Great Amnesia*. Bloomington: Indiana University Press.

CHAPTER NINE

Forests and Trees

I AM convinced," Aldo Leopold wrote in 1941, "that most Americans have no idea what a decent forest looks like. The only way to tell them is to show them" (Leopold, 1991, p. 294). Imagine, however, if we were unable to show them little more than a few acid-drenched, biologically impoverished remnant forests or industrial forests where trees are grown like corn. This is no longer the hypothetical nightmare that it once seemed to be. Excessive logging, corporate monocultures, agriculture, urbanization, road building, recreational development, and air pollution are reducing forested areas around the world and radically undermining the integrity of those that remain. The Amazon is being lost at a rate of approximately 15,000 square kilometers each year (Skole and Tucker, 1993). At present rates of logging, tropical forests of Southeast Asia and Africa will disappear in a matter of decades. Preparations are under way to cut the great boreal forests of Russia. The United States is now engaged in a contentious debate about whether we should save the last 5% to 8% of the ancient forests in the Northwest or preserve one final decade or so of logging jobs. The U.S. Forest Service plans to cut 91,000 of the best 100,000 acres of the Tongass Forest in Alaska. Acid rain continues to work its damage on Appalachian forests and those of central Europe. The unknown effects of global warming and increased ultraviolet radiation loom on the horizon. Overall, the world loses 37,000,000 acres of forest each year (Perlin, 1989, p. 15). Chateaubriand's old lament that forests precede civilization and deserts follow is now a global reality.

More is at stake, however, than the fate of an economic resource. With only pathetic remnants of once majestic forests, how will we instill the *idea* of "decent" forests in the young, whose minds are increasingly warped by television, Nintendo, MTV, shopping malls, freeways, and those sensory deprivation chambers we call suburbs? We are never more than one generation away from losing the idea of forests as places of wildness and ecstasy, mystery and renewal, as well as the knowledge of their importance for human survival. We can now foresee a time, not far off, when no one virtually anywhere will remember the aboriginal forest. But the power behind the idea of decent forests depends on the experience of decent forests, not on secondhand, bookish abstractions.

How should teachers and educational institutions respond to the worldwide decline in forests? There is good reason to believe that the question has not yet been asked in the upper echelons of higher education. Officials at the University of California, Santa Cruz, for example, recently announced plans to sell off 440 acres of forest, which includes 50 acres of old growth redwood trees, some that are more than 700 years old (*San Francisco Chronicle*, April 24, 1993). The land had been donated to the University of California in 1942 by a retired professor and his wife. The news account does not describe the donors' expectations about the use of the land or any restrictions on its resale. The tract is now worth an estimated $1,000,000. Not surprisingly, the decision was motivated by budget cuts imposed by the state of California amounting to over $11,000,000 on the Santa Cruz campus alone. By selling the land, the university can reduce its management costs by $25,000 per year, eliminate any potential liabilities attached to the land, and get a one-time windfall amounting to 9% of its budget deficit.

The news account ends by noting that "the deans of all the academic divisions . . . 'have reviewed the proposal to sell and have found that there is no academic interest to be served by retaining the property.'" As one administrator put it, "the University is an educational institution, not a biodiversity organization." The executive director of the Save-the-Redwoods League, John B. Dewitt, disagrees, saying that the "university should retain the property for the benefit and use of the students and faculty . . . for research and recreation," apparently an outdated view of research, recreation, and institutional purposes.

I know neither the particular land in question, nor the financial details of the case beyond what is contained in the newspaper account. I do know, however, that the story is increasingly familiar. More and more colleges and universities are willing to sell off natural areas in their possession and use the proceeds for what administrators regard as more practical purposes. A few have participated in large-scale commercial developments on university-owned lands. Such actions say more than any number of glossy publications or learned speeches about the real institutional priorities that apparently do not have much to do with trees, forests, and biodiversity. Land holdings, including those in forested land, are appraised mostly for their cash value, not for their value in preserving biodiversity or in educating the young about forests.

Intended or not, decisions to sell off natural lands do have an effect that can be rightly described as educational. Colleges and universities educate by what they do as well as by what they say. Students no doubt will observe that when the going gets a wee bit tough, their intellectual mentors and role models regard natural lands and whatever biological diversity they hold as expendable. They will note that those presuming to educate them rarely see any serious educational value in wild lands or even in ancient trees that contain enough genetic information to fill the college library several times over. Whether outraged or apathetic, they will see that universities, which pay their chief executive officers, various deans, and football coaches six-figure salaries, can seldom find the relatively small amounts of money necessary to manage increasingly rare natural areas. Students may even note the discrepancy in the minute efforts to fund such things as opposed to the effort exerted to fund, say, campus parking facilities. The alert may also correlate such decisions with the manifest decline in their life prospects. All of this provides a liberal education in how educational institutions work and how sometimes they do not work for the good of the world the students will inherit.

The willingness of institutions to preserve natural areas is, however, only a small part of a larger problem. In a culture that regards land as a commodity, it is easy to think of forests and natural areas as merely resources expendable to support other, and more serious, things. This attitude reflects a deep conflict between humankind and forests as old as civilization itself. In *The Epic of Gilgamesh* (circa 4700 B.C., republished in Sandars, 1972), the hero kills the appointed guardian of the forests so that he can "strip the mountains of their cover." By cutting down the cedar forest, he intends to "leave behind . . . an enduring name." The angry gods' reprisal, predictably, takes the form of a series of ecological curses. "It is a sorry fact of history," in Robert Harrison's (1992) words, "that human beings have never ceased reenacting the gesture of Gilgamesh" (p. 18). The epic and its historical reenactment through the ages is a story of violence and madness. The same can be said about the destruction of forests in our time. Forests have been a mirror of sorts, reflecting back whatever humans wished to see. Minds receptive to mystery, like that of John Muir, have seen forests as sacred places. Thoroughly utilitarian minds see little more than something to sell, whether board feet or visitor days. University administrators hard pressed for cash see windfall profits. But whatever our rationalizations and practical material needs, "there is too often," in Harrison's words, "a deliberate rage and vengefulness at work in the assault on nature and its species" (p. 18).

In the coming century humankind will need healthy forests more than ever, both for practical reasons of survival and to preserve sanity, an utterly practical reason. The effort to rethink ancient habits and antagonisms must begin with a different kind of education in which forests become a significant part of the general curriculum. That will require confronting paradoxes deeply embedded in the curriculum about forests and their relation to human progress.

Madav Gadgil and Ramachandra Guha (1992) describe two such paradoxes, the first of which has to do with the fact that

> hunter-gatherers live *in* in the forest, agriculturalists live adjacent to but within *striking distance* of the forest, and urban-industrial men live *away* from the forest. Paradoxically, the more the spatial separation from the forest the greater the impact on its ecology, and the further removed the actors from the consequences of this impact! (p. 52)

In other words, forests out of sight are forests out of mind. Civilization was planted in a clearing both literally and figuratively. As civilization grew, forests receded, becoming ever more peripheral to our imagination and to our sense of reality. Our use of forests, accordingly, has become increasingly mindless, driven by large impersonal forces that undermine our long-term prospects.

The second paradox is even more threatening to our established academic ways:

> The faster the development of formal, scientific knowledge about the composition and functioning of forest types, the faster the rate of deforestation. . . . the belief that science provides an infallible guide has nonetheless encouraged major interventions in natural ecosystems, and these have had unanticipated and usually unfortunate consequences. The history of both fisheries and forest management are replete with illustrations of the failure of sustained-yield methods to

forestall ecological collapse. . . . *Religion and custom as ideologies of resource use are perhaps better adapted to deal with a situation of imperfect knowledge than a supposedly 'scientific' resource management* [emphasis added]. (Gadgil and Guha, 1992, p. 52)

In other words, in dealing with complex systems about which we know relatively little, humility that leaves a large margin for error is smart over the long haul. "The forest," according to Richard Manning (1992), "is a wonder beyond our comprehension" (p. 171). Much of it will remain beyond our comprehension, science notwithstanding. Accordingly, we will need a different manner of thinking about forests that acknowledges forthrightly the limits of our knowledge and our inconsistency in using what knowledge we do have.

The observations of Gadgil and Guha have significant implications for education. Resolution of their first paradox requires bringing trees to the forefront of our consciousness. This can be done first by changing the philosophy of landscape management in which trees on the campus are regarded as little more than decoration. Most colleges and universities intend their campuses to look like country clubs, weedless and biologically sterile places maintained by an unholy array of chemicals. Campus landscapes ought to be more imaginatively designed to promote biological diversity and ecological resilience and to raise the collective ecological IQ of the campus community. Campuses ought to be maintained as natural areas that harbor biological diversity. The institutional calendar might also include annual celebrations around tree planting and landscape restoration. Who knows, perhaps even administrators in vulnerable moments might be persuaded to take a walk in a woods!

Resolution of Gadgil and Guha's second paradox will require the integration of forests throughout the curriculum so that all students know beyond any shadow of a doubt how their prospects are intertwined with those of forests. A generation that will need the ecological services of forests more than any previous generation will need a deeper comprehension of how forests work (Maser, 1989). They will need better political and social mechanisms to protect forests, now vulnerable to the tragedy of the commons on a global scale. They will need an honest economics that accounts for all the values of forests (O'Toole, 1988; Panayotou and Ashton, 1992; Repetto and Gillis, 1988). They will need to know the historical relationship between forests and their own history (Williams, 1989). They will need to know a great deal about the practical uses of trees in working landscapes (Smith, 1988). They will need to understand the relationship between forests and the evolution of the human mind (Harrison, 1992). They will need a larger idea of forests and wildness than that contained in the industrial worldview (Oelschlaeger, 1991). They will also need stories and myths that give purpose and meaning to the experience of forests (Giono, 1985). And they will need the example of mentors willing to fight for every tree, woods, scrap of remaining wildness, and decent forest.

SOURCES

Giono, J. 1985. *The Man Who Planted Trees*. Post Mills, VT: Chelsea Green Publishing Co.

Gadgil, M., and Guha, R. 1992. *This Fissured Land: An Ecological History of India*. Berkeley: University of California Press.

Harrison, R. 1992. *Forests: The Shadow of Civilization*. Chicago: University of Chicago Press.

Leopold, A. 1991. *The River of the Mother of God and Other Essays by Aldo Leopold*. S. Flader and J. B. Callicott, eds. Madison: University of Wisconsin Press. (Original work published 1941.)

Manning, R. 1992. *Last Stand*. New York: Penguin Books.

Maser, C. 1989. *Forest Primeval*. San Francisco: Sierra Club Books.

Oelschlaeger, M. 1991. *The Idea of Wilderness*. New Haven: Yale University Press.

O'Toole, R. 1988. *Reforming the Forest Service*. Washington, DC: Island Press.

Panayotou, T., and Ashton, P. 1992. *Not by Timber Alone: Economics and Ecology for Sustaining Tropical Forestry*. Washington, DC: Island Press.

Perlin, J. 1989. *A Forest Journey: The Role of Wood in the Development of Civilization*. New York: Norton.

Repetto, R., and Gillis, M., eds. 1988. *Public Policies and the Misuse of Forest Resources*. New York: Cambridge University Press.

Sandars, N. K., ed. 1972. *The Epic of Gilgamesh*. New York: Penguin Books. (Original work circa 4700 B.C.)

Skole, D., and Tucker, C. 1993. Tropical Deforestation and Habitat Fragmentation in the Amazon. *Science*, 260, p.p. 1905–1910.

Smith, J. R. 1987. *Tree Crops: A Permanent Agriculture*. Washington, DC: Island Press.

Williams, M. 1989. *Americans and Their Forests*. New York: Cambridge University Press.

CHAPTER TEN

Politics

PROPERLY speaking, there is no "crisis of biological diversity" or even an "ecological crisis." But there is a large and growing political crisis that has ecological and other consequences. The primary causes of biotic impoverishment are not ignorance or the lack of research funding. They are, on the contrary, invariably political, having to do with "who gets what, when, and how." The decisions necessary to conserve biological diversity likewise will be political. To describe the present breakdown of collective and individual judgment and the atrophy of leadership, public purposes, and foresight as an ecological crisis implies that the problem is the lack of knowledge about ecology and that the solution therefore is more research, more research funding, more publications, more conferences, and so forth. As important as research is, the lack of it is not the limiting factor in the conservation of biological diversity. For all that we do not know, we know without question that we are rapidly unraveling ecosystems and destabilizing the biosphere with consequences that cannot be good, which is to say that we know enough to act far more prudently, to conserve, and to restore that which through carelessness we have destroyed and are destroying. Why then do we find it so difficult to do what is merely obvious and necessary?

I propose four answers. First, we have defined the problem wrongly as one of science, not one of politics. Accordingly, we have focused on the symptoms and not the causes of biotic impoverishment. The former have to do with the vital signs of the planet. The latter have to do with the distribution of wealth, land ownership, greed, the organization of power, and the conduct of the public business. These are large, complex, and to some, disagreeable subjects, and there are unspoken taboos against talking seriously about the very forces that undermine biological diversity. I am referring to our inability to question economic growth, the distribution of wealth, capital mobility, population growth, and the scale and purposes of technology. These subjects have not yet entered the public dialogue because they are not considered realistic. But until they do, we are not likely to conserve much.

Second, the conservation of biological diversity is difficult because the generally anemic state of democracy here and elsewhere does not favor the conservation of much of anything. In the face of mounting crises, the language of politics has become confused, convoluted, and timid and a good indicator of bamboozlement at work. For evidence, consider the inability of high officials to describe our involvement in the war over Kuwait as one having to do with our addiction to cheap oil and the lack of a national policy to promote energy efficiency. Americans were willing to risk the lives of the young—both American and Iraqi—to ensure our access to cheap oil, while refusing to use American ingenuity to minimize our need for it. Our inability then and now to make energy efficiency a national goal, for reasons having to do with good economics, prudent ecology, and real national security is a massive political and moral failure. I might just as well have cited our inability to confront what has become a permanent budget crisis or the theft of hundreds of billions from savings and loan institutions.

Third, there is in America and perhaps western Europe the remarkable belief that we won the Cold War. It is certainly clear that communism has failed, but that does not mean that we "won." We are mired in a deeper crisis that has to do with many of the core values and assumptions of the modern world. Vaclav Havel (1990) described it as the "process of anonymization and depersonalization of power and its reduction to a mere technology of rule and manipulation" (p. 143). The answer according to Havel is an

> 'anti-political politics,' that is, politics not as the technology of power and manipulation, of cybernetic rule over humans or as the art of the useful, but politics as one of the ways of seeking and achieving meaningful lives, of protecting them and serving them. (p. 155)

The triumph of instrumental rationality, Havel argued, is no less characteristic of western nations than of those formerly behind the iron curtain. Instead of leadership and democratic participation, our politics are the product of slick packaging, public relations managers, and 30-second television commercials. We are losing our ability to discuss the things we have in common, including our common dependence on the biosphere. The answer Havel has proposed is to

> reconstitute the natural world as the true terrain of politics. . . . We must draw our standards from our natural world, heedless of ridicule, and reaffirm its denied validity. We must honour with the humility of the wise the bounds of that natural world and the mystery which lies beyond them admitting that there is something in the order of being which evidently exceeds all our competence. (p. 149)

Finally, the large gap between strong public support for the environment and the environment as a potent national political issue is partly explained, I think, by the failure of scientists to communicate adequately to society. The role of science as a political force is still limited by what Thomas Kuhn once called the paradigm of "normal science." Scientists tend mostly to talk to other scientists and not often enough to the public or to its elected leaders. There is, some argue, a professional bias against scientists communicating to the public in the common language. The workings of normal science and peer pressures may even tend to encourage timidity and caution as a scientist attempts to appear "reasonable" and "objective." The result, in George Woodwell's (1989) words, is "hyperobjectivity," which "is the epitome of unreasonableness, and its practice not only lends support to avarice and pollution, but destroys the credibility of science and scientists as a source of simple common sense" (p. 15). It is now time for scientists to speak more clearly and boldly about larger risks, as George Woodwell, Rachel Carson, Paul Ehrlich, James Hansen, and Tom Lovejoy have done.

What does this have to do with education? The short answer is "everything." Ecological education is not just about biology, it is equally about the deeper causes of biotic impoverishment, which have to do in one way or another with political behavior, institutions, and philosophies. Conservation biology is a dialogue between science and political action. The fate of the biosphere depends on how we answer questions about

- the possibilities for international institutions,
- the role and function of national governments,
- the appropriate degree of political centralization,
- the scale of technology,

- constraints on capital,
- the distribution of land and wealth, and
- the future of democratic participation.

Answers to these questions will grow more divisive in coming decades and pressures will mount to find short-term technical fixes to problems that are essentially political or to continue to deny the problems altogether. In addition, we need to make it easier for scientists to speak out and to become a force for public enlightenment without losing research funds or being denied tenure or promotion. We need to allow risk taking and honor those willing to do so. Finally, it is time to establish national goals for ecological literacy and make these a vital part of the curriculum of public schools and colleges. By ecological literacy, I mean an understanding of both the biology of conservation and the political basis of conserving societies.

SOURCES

Havel, V. 1990. *Living in Truth*. London: Faber & Faber.
Woodwell, G. 1989, February. On Causes of Biotic Impoverishment. *Ecology, 70, 1.*

CHAPTER ELEVEN

Economics

IMAGINE being in the cockpit of a 747 jet in which two pilots were in disagreement about whether the plane faced imminent disaster. Passengers and crew of even modest good sense would deem it a matter of the highest priority to determine which pilot was right. The debate would most likely appear to them as one transcending merely academic interest. Both pilots might be correctly reading different dials. One could be looking, say, at the altimeter, the on-board radar, and the position of the wing flaps, while the other is reading the fuel gauge, air speed indicator, and cabin pressure dial.

The world faces a somewhat analogous situation. To the biologist concerned about conservation, the dials and gauges reporting on the state of the world indicate potentially catastrophic rates of species loss, a sharp decline in major biomes such as coral reefs, wetlands, and rain forests, overpopulation pointing to even greater stresses in the future, high rates of soil loss and land abuse, and perhaps the onset of global warming. Humanity now uses or co-opts 40% of terrestrial net primary productivity, leading many biologists to see a collision ahead between population and economic growth and the carrying capacity of the earth.

The dials and gauges that economists read, in contrast, give them reason for optimism. Gross world product has increased throughout the twentieth century by some 1300% and continues to rise. Per capita wealth among all developed countries continues to grow. Most important, technological innovation continues to improve energy and resource efficiency. Moreover, some believe that technology and higher prices will combine to create substitutes for scarce resources (Dasgupta, 1991, pp. 107–126).

From their perspective, the essential problem is that of getting the price of things right and letting the market and technology do the rest.

Biologists, in other words, are paying attention to the larger economy of life: the biosphere; economists are looking at the subeconomy humans have built by exploiting nature. Both are important, but not equally so. The larger economy is nearly everywhere showing signs of stress and breakdown while the latter is still expanding. The question is not which indicators are more accurate but which is more basic and over what time span. This is a wager of sorts over the extent to which technology can render human economies independent of healthy soils, microbes, plants, animals, forests, ecosystems, and a stable climate. The biologist bets on bugs and biota, the economist on prices and technology. Beneath this, another wager is being played out over the extent to which technology can separate humanity from nature emotionally, spiritually, and intellectually, and whether the person in this Brave New World may regress to some frightful subhuman level.

Those who believe that such wagers cannot be won and should not be made make two basic arguments against mainstream neoclassical economics. The first has to do with the failure of economists to include the price of natural services in their calculations of welfare and income. Harvard biologist E. O. Wilson (1989), for example, argued that

> neoclassical economics is bankrupt. Its quantitative models of optimization and equilibrium have no realistic measure to place on the value of the environment. Economists cannot factor in opportunity costs, the losses incurred when habitats are destroyed and species go extinct. They are unable to handle multiple margins outside a narrowly defined market economy. (p. 7)

In other words, as biotic stocks, such as forests, soils, and wildlife, are destroyed, their loss should be subtracted from measures such as gross national product in the same manner as capital depreciation is subtracted from corporate profit and loss statements. The result would be a net figure showing the loss of future productivity owing to the loss of natural capital. This is altogether sensible and requires no major disciplinary shift beyond a commitment to full and honest accounting that includes biotic impoverishment. To make such calculations, however, one must first know a great deal about the services that various parts of nature provide directly or indirectly. No prudent ecologist, however, would claim that we now have or will soon have knowledge this extensive. Moreover, one must make a simplifying assumption that those parts of nature deemed useless have no value beyond the arbitrary and fickle public "willingness to pay." No worthy philosopher would be so bold without first writing several volumes of caveats.

World Bank economist Herman Daly has made a second and more radical critique. Mainstream economists regard the economy, in Daly's view, as an isolated, closed system with no exchange of matter or energy with the environment. As a result, economists working within this framework are unable to deduce any optimal scale for the overall economy. Daly (1991) proposed instead a model of the macroeconomy as

> an open subsystem of the ecosystem . . . dependent upon it, both as a source for inputs of low-entropy matter-energy and as a sink for outputs of high-entropy matter-energy. (p. 256)

In other words, physics and ecology are more basic than economic theory. From the perspective of physics or ecology the idea of an ever expanding economy within a biosphere of fixed size is, in Lewis Thomas's (1984) view, "stupidity on the grandest scale." But rates of stupidity among economists fully committed to the ever expanding economy are surely no higher than those found, say, among members of the U.S. Congress or the British Parliament. Aside from the power of religious faith disguised as technological optimism, how is the paradox of manifestly clever people believing the truthfulness of physical impossibilities to be explained? Partial answers can be found in assumptions economists bring to their work, which predispose them not to see the limits of natural systems that are the daily stuff of biology. Beyond such answers, however, one encounters a logic that goes like this: (a) Everyone wants what we, the rich, have, which is to say wealth; (b) who are we to say they should not have it?; (c) therefore, we cannot deny "progress" and the human desire for material improvement. Not much is said of the roughly $450 billion spent worldwide on advertising each year to manufacture wants. But that quibble aside, the unstated assumption is that we can summon neither the civic and moral wisdom to create a more equitable distribution of wealth nor the wit to redefine well-being in a less stuff-oriented and ecologically destructive manner.

Such logic traps its adherents into accepting potentially catastrophic risks that sanity and prudence suggest we should avoid. The inherent risks, for example, in nuclear power, genetic engineering, and nanotechnologies appear "necessary" only because we have trapped ourselves in yet another crisis of carrying capacity. And at each level, the stakes ratchet upward.

One might dismiss the discrepancy between economists and biologists were it not for the fact that the assumptions of neoclassical economics have become widely and often uncritically accepted as faith and dogma so that the economists' model of "economic human" serves as both an adequate description of human behavior, which it is not, and as a prescription of how rational beings ought to behave, which is to say, selfishly. Moreover, capitalism and its economists are on the march. After the fall of communism, armies of economists and business school professors headed eastward to show former communists how to operate their fallen economies. As a result the ecological flaws of our economic theories have taken on added importance in the affairs of more people over a larger share of the planet.

Those same theories so confidently expounded and exported are a major cause of the crisis of biodiversity because they assign only short-term monetary value to species, ecosystems, climate stability, and the well-being of future generations. The problem in its larger context is what value should be given to (a) those life forms, landscapes, and ecological processes whose value cannot or should not be stated in monetary terms and hence cannot be appropriately regarded merely as resources; (b) those whose value is unknown and perhaps unknowable; and (c) those, now ill-considered, that we may yet come to appreciate. Biologists and philosophers can help identify those parts of natural systems and natural areas that fit Category a. By simply acknowledging their ignorance and the limits of science, they can also sound a cautionary alarm with regard to Category b. The third category, however, is more troublesome because it transcends biology and economics. It has to do with our maturity as a species, by which I mean our capacity to identify with the biotic community and to shelter life. However, we may learn someday to value nature beyond the wildest dreams of present-day economists. At least we should hold out the possibility, and doing so may even help us to mature a bit.

SOURCES

Daly, H. 1991. Towards an Environmental Macroeconomics. *Land Economics*, 67, pp. 255–259.

Dasgupta, P. 1991. Exhaustible Resources. In L. Friday and R. Laskey, eds., *The Fragile Environment*. New York: Cambridge University Press.

Repetto, R., and Gillis, M., eds. 1988. *Public Policies and the Misuse of Forest Resources*. New York: Cambridge University Press.

Thomas, L. 1984. Scientific Frontiers and National Frontiers: A Look Ahead. *Foreign Affairs* 62, pp. 966–994.

Wilson, E. 1989. Conservation: The Next Hundred Years. In D. Western and M. Pearl, eds., *Conservation for the Twenty-First Century*. New York: Oxford University Press.

CHAPTER TWELVE

Judgment: Pascal's Wager and Economics in a Hotter Time

IN WEIGHING the question concerning the existence of God, seventeenth-century philosopher and mathematician Blaise Pascal (1941) proceeded in a manner perhaps instructive for other and more mundane questions. "Reason," he declared, "can decide nothing here." Nonetheless, "you must wager. It is not optional." You have, he believed,

> two things to lose, the true and the good; and two things to stake, your reason and your will, your knowledge and your happiness; and your nature has two things to shun, error and misery.

What would you lose by believing that God exists and living a life accordingly? Pascal's answer was "If you gain, you gain all; if you lose, you lose nothing." By doing so you would become "faithful, honest, humble, grateful, generous, a sincere friend, truthful." The opposite decision that God did not exist and a life lived in pursuit of "poisonous pleasures, glory and luxury," whatever its short-term gains, would be one of misery. In other words, if you chose not to believe and it turned out that God did exist, you would have hell to pay. On the other hand, if God did not exist and you had lived a life of faith you would have sacrificed only a few fleeting pleasures but gained much more. Pascal's argument for faith, then, rested on the sturdy foundation of prudential self-interest aimed to minimize risk.

The world now faces a somewhat analogous choice. On one side a large number of scientists believe that the planet is warming rapidly. If we continue to spew out heat-trapping gases, such as carbon dioxide, methane, chlorofluorocarbons, and nitrous oxide, these scientists say we will warm the planet intolerably within the next century. The consequences of dereliction and procrastination may include killer heat waves, drought, sea-level rise, superstorms, vast changes in forest and biota, considerable economic dislocation, and increases in disease: a passable description of hell. But like Pascal's wager, no one can say with absolute certainty what will happen until the consequences of our choice, whatever they may be, are upon us. Nonetheless, "we must wager. It is not optional."

Others, however, claim to have looked over the brink and have decided that hell may not be so bad after all, or at least that we should research the matter further. Yale University economist William Nordhaus (1990b), for example, believes that a hotter climate will mostly affect "those sectors [of the economy] that interact with unmanaged ecosystems" such as agriculture, forestry, and coastal activities." The rest of the economy, including that which operates in what Nordhaus (1990b) called "a carefully controlled environment," which includes shopping malls and presumably the activities of economists, will barely notice that things are considerably hotter. "The main factor to recognize," Nordhaus (1990a) asserted, "is that the climate has little economic impact upon advanced industrial societies" (p. 193).

Nordhaus concluded that "approximately 3 percent of U.S. national output originates in climate-sensitive sectors and another 10 percent in sectors modestly sensitive to climatic change." There may even be, he noted, beneficial side effects of global warning: "The forest products industry may also benefit from CO_2 fertilization." (It is, I think, no mistake that he did not say "forest" but rather "forest products industry.") Construction, he thinks, will be "favorably affected" as will "investments in water skiing." In sum, Nordhaus's "best guess" is that the impact of a doubling of carbon dioxide "is likely to be around one-fourth of 1 percent of national income," an estimate that he confessed has a "large margin of error" (p. 195).

Nordhaus (1990b), however, wishes not to be thought to favor climate change. Rather, the point he tried to make is that "those who paint a bleak picture of desert Earth devoid of fruitful economic activity may be exaggerating the injuries and neglecting the benefits of climate change" (p. 196). Whether a hotter earth, but one not "devoid of fruitful economic activity" might, however, be devoid of poetry, laughter, sidewalk cafes, forests, or even economists he does not say. But he did note that there are a number of technological responses to our plight, including "climate engineering . . . shooting particulate matter [books on economics?] into the stratosphere to cool the earth or changing cultivation patterns in agriculture." Nordhaus, an economist, gave no estimate of the costs, benefits, or even feasibility of these "options." He did, however, estimate the cost of reducing carbon dioxide emissions by 50% as $180 billion per year. Faced with such costs, Nordhaus expressed the view that "societies may choose to adapt," which in his words means "population migration, capital relocation, land reclamation, and technological change" (1990b, p. 201), solutions for which he again has given no cost estimate. What about those who cannot adapt, migrate, buy expensive remedies, or relocate their capital? Nordhaus does not say, and one suspects that he does not say because he has not thought much about it.

The complications Nordhaus (1990b) has noticed have to do with "how to discount future costs and how to allow for uncertainty." A discount rate of, say, 8% or higher would lead us to do nothing about warming for a few decades while the problem grows gradually or perhaps rapidly worse. A rate of 4% or less "would give considerable weight today to climate changes in the late twenty-first century." What is Nordhaus's solution? "The efficient policy," he argued, "would be to invest heavily in high-return capital now and then use the fruits of those investments to slow climate change in the future" (p. 205). He described this as a "sensible compromise" between what he asserts is a "*need* for economic growth" and "the *desire* for environmental protection" [emphasis added], that is, one more binge, virtue later.

To his credit, Nordhaus (1990b) has acknowledged that "most climatologists think that the chance of unpleasant surprises rises as the magnitude and pace of climatic change increases" (p. 206). He has also noted that the discovery of the ozone hole came as a "complete surprise," suggesting the possibility of more surprises ahead. But in the end he has come down firmly in favor of what he calls "modest steps" that "avoid any precipitous and ill-designed actions that [we] may later regret," actions that he does not specify, making it impossible to know whether they would be in fact precipitous, ill-designed, or regrettable. Nordhaus has stated the belief that "reducing the risks of climatic change is a worthwhile objective" but one in his opinion not more important than "factories and equipment, training and education, health and hospitals, transportation and communications, research and development, housing and environmental protection" (p. 209) and so forth. He seems not to have noticed the close relationship between heat, drought, and climate instability, on one hand, and the economy, public health, human behavior under stress, and even what he has called "environmental protection," on the other.

One might dismiss Nordhaus's analysis as an aberration were it not characteristic of the recklessness masquerading as caution that prevails in the highest levels of government and business here and elsewhere and were he not as influential at these levels as he certainly is. Nordhaus's views on global warming are neither an aberration within his profession nor are they without consequence where portentous choices are made.

Nordhaus's opinions about global warming, for example, weighed heavily in the 1991 report issued by the Adaptation Panel of the National Academy of Sciences (1991). The panel, which included Nordhaus, approached global warming as an investment problem requiring the proper discount rate. However, for those whose interests were discounted, such as the poor and future generations, the problem appears differently, as one of power and intergenerational responsibilities. The panel, moreover, assumed a great deal about the adaptability of complex, mass, technological societies under what may be extreme conditions. In citing "the proven adaptability of farmers," for example, are they referring to the 4 million failed farms in the past 50 years? Or to those 1.5 million farms presently at or close to the margin? Or are they referring to the overdependence of agriculture and food distribution systems on the very fossil energy sources that are now heating the earth? Or perhaps to present rates of soil loss and groundwater depletion due to current farm practices? Can farmers adapt if warming is sudden? Since people live "in both Riyadh and Barrow," the panel drew the implication that humans are almost infinitely adaptable, while admitting that some cities will have to be abandoned and people in poorer countries may be substantially harmed. The panel smartly hedged its bets by admitting that the warming could be sudden and catastrophic but quickly dismissed these possibilities. They did not ask what could happen beyond their 50-year horizon, nor did they ask about the effects on American society of making such portentous decisions in the same way that investment decisions are made about building bridges or shopping malls.

It is therefore a matter of concern that such analysis gives considerable aid and comfort to those with all too much to gain by ignoring the risks involved in climate change or the benefits of a farsighted energy policy. Accordingly, we should attempt to understand how such thought comes to pass, whose ends it serves, and what consequences it risks.

By comparison, it is instructive to note that atmospheric physicists, climate experts, and biologists agree almost without exception that the theory of global warming is beyond dispute (Intergovernmental Panel on Climate Change, 1991). It is widely agreed that heat-trapping gases in the atmosphere do in fact trap heat. If we put enough of these in the atmosphere, we will trap a great deal of heat. There is further agreement that if the warming turns out to be rapid, the consequences would in all probability be widely catastrophic even though we cannot predict these with absolute certainty. Disagreement focuses on matters having to do with rates, thresholds, and the effects of feedbacks that might enhance or retard rates of warming. However these are decided, there is no doubt whatsoever that by increasing heat-trapping gases to levels higher than any in the past 160,000 years and at rates far more rapid than characteristic of past climate shifts, we are conducting an unprecedented experiment with the earth and its biota. This experiment need not, and should not, be carried out. But like Pascal's wager, certainty about the consequences will come only after all bets are called in.

Given what is at stake, errors of fact and logic committed by Nordhaus and the Adaptation Panel deserve close attention. For example, the belief that decline in agriculture and forestry would be of little consequence because they are only 3% of the U.S. economy is equivalent to believing that since the heart is only 1% to 2% of bodyweight it can be removed or damaged without consequences for one's health. Both Nordhaus and the Adaptation Panel regard the economy as linear and additive without straws that break the back of the camel, surprises, thresholds of catastrophe, or even places where angels would fear to go. The biological facts underlying the research are also suspect. There are many reasons to believe that "CO_2 fertilization" will not enhance farm and forest productivity as Nordhaus and the Adaptation Panel believe. Changes in rainfall, temperature, and biological conditions would more likely reduce growth (Houghton and Woodwell, 1989). Higher temperatures mean higher rates of respiration, hence the release of still more carbon and methane. The rate of climate change may well be many times faster than that to which plants and animals can adapt. This will mean at some unknown date a dieback of forests and the release of even more carbon through fire and rapid decay. It will also mean a sharp reduction in biological diversity.

Economic estimates used by Nordhaus and the Adaptation Panel are also questionable. Both ignore a large and growing body of evidence that the actions necessary to minimize global warming would be good for the economy, human health, and the land. Studies by the U.S. Environmental Protection Agency, the Electric Power Research Institute, and independent researchers all point to the same conclusion: energy efficiency, which reduces the emission of carbon dioxide, is not only inexpensive, it is in fact a prerequisite of economic vitality. The U.S. economy is roughly one-half as energy efficient as that of the Japanese. This fact translates into a 5% cost *disadvantage* for comparable U.S. goods and services (Lovins, 1990). Instead of an annual cost that Lovins estimates at $180 billion, more reliable studies have shown a net savings of approximately $200 billion from improvements in energy efficiency. This, in Lovins's words, is not a free lunch, but a lunch we are paid to eat. However, estimates by Nordhaus or the Academy Panel do not include the costs of relocating millions of people, the costs of failing to do so, the costs entailed in diking coasts, the costs of international conflicts over water, the costs of importing food when the plains states become drier, or the costs of changes in diseases due to climate change. Nor does Nordhaus or the Academy Panel say what the cost might be if global warming turns out to be rapid and full of even worse surprises.

The practice of discounting the future creates other costs that cannot be quantified but that will be assessed. If they had included the preferences of, say, the third generation hence in the equation, their conclusions would have been quite different. Nordhaus and the Academy Panel chose not to do so, however, by assuming that investments in more of the same kinds of activities that have created the problem in the first place are "worthy goals." On closer examination, most of these will intensify the problem of global warming and dig us in still deeper while ignoring opportunities to invest in energy efficiency and renewables that would reduce the emission of heat-trapping gases in an economically sound manner.

The economic estimates of Nordhaus and the Academy Panel are not to be trusted because their economy is an abstraction independent of bio-physical realities, comparable, say, to an airline pilot who regarded the law of gravity as merely an interesting but untested theory. Their economics are not to be trusted because they fail to acknowledge the vast and unknowable complexity of planetary systems, which cannot be "fixed" by any technology without courting other risks. Their economics cannot be trusted because they are not very good economics. They have ignored the relationship between economic prosperity and energy efficiency, as well as that between energy efficiency and the emission of greenhouse gases. Their economics are not to be trusted because the problem of global warming is not first and foremost one of economics as they believe, but rather one of judgment, wisdom, and love for the creation.

Their economics cannot be trusted because they do not include flesh-and-blood people who, under conditions of a rapidly changing climate, will not act with the rationality presumed in abstract models concocted in air-conditioned rooms. Real people stressed by heat, drought, economic decline, and perhaps worse will curse and kill more often and celebrate and love less often. And they will mourn the loss of places disfigured by heat, drought, and death that were once familiar, restoring, and consoling.

Finally, the economics of Nordhaus and the Adaptation Panel cannot be trusted because they would have us risk this and more for another decade or two of business as usual, which as we now know does not mean sustainable prosperity or basic fairness. This is a foolish risk for reasons Pascal described well. If it turns out that global warming would have been severe and we forestalled it by becoming more energy efficient and making a successful transition to renewable energy, we will have avoided disaster. If, however, it turns out that factors as yet unknown minimized the severity and impact of warming while we became more energy efficient in the belief that it might be otherwise, we will not have saved the planet, but we will have reduced acid rain, improved air quality, decreased oil spills, reduced the amount of strip mining, reduced our dependence on imported oil and thereby improved our balance of payments, become more technologically adept, and improved our economic competitiveness. In either case we will have set an instructive and far-sighted precedent for our descendants and for the future of the earth. If we gain, we gain all; if we "lose," we still gain a great deal. On the other hand, if we do as Nordhaus and the members of the Adaptation Panel would have us do, and the warming proves to be rapid, there will be hell to pay.

SOURCES

Houghton, R. A., and Woodwell, G. 1989, April. Global Climatic Change. *Scientific American*, pp. 36–44.

Intergovernmental Panel on Climate Change. 1991. *Climate Change*. New York: Oxford University Press.

Lovins, A. 1990. The Role of Energy Efficiency. In J. Leggett, ed., *Global Warming*. New York: Oxford University Press.

National Academy of Sciences. 1991. *Policy Implications of Greenhouse Warming: Report of the Adaptation Panel*. Washington, DC: National Academy Press.

Nordhaus, W. D. 1990a, July 7. Greenhouse Economics: Count Before You Leap. *The Economist*.

Nordhaus, W. D. 1990b. Global Warming: Slowing the Greenhouse Express. In H. Aaron, ed., *Setting National Priorities*. Washington, DC: Brookings Institution.

Pascal, B. 1941. *Pensées*. New York: The Modern Library.

PART THREE

RETHINKING EDUCATION

IN RELATION to our long-term prospects on earth, by what specific standards do we judge schools, colleges, and universities? If our intent in such places is to equip people to be citizens of the biotic community, what will they need to know and how should they learn it? The essays in Part three were written in the belief that the standard measures for educational quality will have to be changed to account for how institutions and their graduates affect the biotic world. Accordingly, the chapters deal with matters of institutional standards, the disciplinary organization of learning, curriculum, and academic architecture.

CHAPTER THIRTEEN

Rating Colleges

O NE OF the more consistent idiosyncracies of Americans is their penchant for ranking things. It is, on the whole, a harmless pastime, giving indoor pleasure to many, and bestowing high status upon those called on to create and maintain various rankings. It has also been known to boost sales of publications of one sort or another and, like *Sports Illustrated*'s annual swimsuit issue, it provides either agreeable diversion or a source for moral indignation during an otherwise dull part of the year. One should not presume, however, that the relationship between such lists and reality is great. Nor is it necessary that it be so. Their function, rather, is to gratify, amuse, employ, sell, or fuel disagreement, hence the development of subsequent lists and rankings.

Until recently, colleges and universities, for the most part, ranked themselves. After due consideration the great majority solemnly proclaimed themselves to be "excellent." But hundreds, if not thousands, of institutions laying claim to an attribute scarce by definition gives scant basis for ranking. The subsequent loss of pleasurable contention has been considerable. We have been rescued from this plight by various guides to colleges, including those by *The New York Times* and *U.S. News and World Report*. These and others like them rank colleges on such things as peer reputation, Scholastic Aptitude Test scores of incoming freshmen, size of endowments, number of books in the library, percentage of PhDs on the faculty, publications by faculty, tuition, faculty–student ratios, and so forth. These purport to describe, in one way or another, the capacity of educational institutions to educate.

Educational institutions, however, are not like football teams, so judging the capacity of a college to cultivate the higher qualities of life and mind is considerably more subtle and complex than appraising the ability of 11 men to do mayhem for 60 minutes. Good education, in fact, may be inversely proportional to many of the qualities now used to rank educational institutions. Peer reputation may be an index only of snobbery and pomposity. Faculty publications may be an indicator of student dissatisfaction and the decline of forests. Large endowments might be a reasonable index of institutional torpor. Research grants may, on occasion, reflect ties to corporate and U.S. Department of Defense activities that Boards of Trustees might rather conceal.

Ranking works best when things are simple and can be easily counted. But good educational institutions are complex, creative, and difficult to describe in numbers. This is why I think that the editors of *U.S. News and World Report*'s college issue would have ranked Plato's Academy rather far down on its list of "regional" institutions. Its library by all accounts was small, it had no laboratories, its student body consisted mostly of locals, and the major professor of the founder, whose work has descended to us by hearsay, was highly discredited through a lifetime of rabble rousing and carousing among the city's young.

There is yet a second problem. Most ranking systems face backwards, using measures that no longer describe present realities or the role of the institution in relation to those realities. For example, whatever their stated purposes, colleges and universities have played a major role in the industrialization of the world in the belief that the domination of nature, on balance, was a good thing. The reality, however, has changed. We have several centuries of hard work ahead of us to clean up the mess: sequestering toxic and radioactive wastes; restoring depleted and mined land; cleaning up lakes, seas, and rivers; stabilizing climate; replanting forests; protecting whatever biological diversity we can; rebuilding decayed urban areas; and bringing all of the other vital signs of earth back to health.

Accordingly, I propose a different ranking system for colleges based on whether or not the institution and its graduates move the world in more sustainable directions. Does four years at a particular institution instill knowledge, love, and competence toward the natural world or indifference and ignorance? Are the graduates of this or that college suited for a responsible life on a planet with a biosphere? This is an admittedly difficult, but not impossible, task. I propose that colleges and universities be ranked on the basis of five criteria.

The first of these has to do with how much of various things the institution consumes or discards per student. Arguably, the best indicator of institutional impacts on the sustainability of the earth is how much carbon dioxide it releases per student per year from electrical generation, heating, and direct fuel purchases. Other ratios of interest would include amounts of paper, water, materials, and electricity consumed per student. These can only be determined by careful audits of how much of what enters and leaves the campus (Smith, 1992). On this basis colleges might compete to become increasingly efficient in lowering resource use per student.

A second basis for ranking has to do with the institution's management policies for materials, waste, recycling, purchasing, landscaping, energy use, and building. What priority does the institution give to the use of recycled materials? What percentage of its material flows are recycled? Does it limit the use of toxic chemicals on the grounds and in buildings? Does it emphasize energy efficiency and solar energy in renovations and new buildings? Does it use nontoxic materials?

Third, does the curriculum provide the essential tools for ecological literacy? What percentage of its graduates know the rudiments of ecology? Do they understand that no good economy can be built on the ruins of natural systems? Do they have experience in the out-of-doors? Is there opportunity and encouragement to restore some part of the nearby rivers, prairies, worn-out farmland, or strip-mined land? Do they understand the rudiments of environmental ethics? Do they understand the difference between optimum and maximum, stocks and flows, design and planning, renewable and nonrenewable, dwelling and residing, sufficiency and efficiency, can do and should do, health and disease, development and growth, and intelligence and cleverness? This presumes, of course, that the faculty itself is ecologically literate and relates environmental themes to course material.

My fourth criterion has to do with institutional finances. Does the institution use its buying power to help build sustainable regional economies? What percentage of its food purchases come from nearby farmers? In studies of food buying at Hendrix College, Oberlin College, Saint Olaf College, and Carleton College, for example, students discovered significant opportunities to increase food quality, decrease costs, and help the local economy. The same approach could be applied throughout all institutional purchases, giving priority to local craftspeople, merchants, and suppliers. Use of institutional buying power to help rebuild local and regional economies is also a prudent hedge against future price shocks associated with higher energy costs coming from supply interruptions, future scarcity, and the eventual imposition of carbon taxes to reduce emission of greenhouse gases.

Colleges and universities also have investment power. To what extent are their funds invested in enterprises that move the world toward sustainability? All institutions should set long-term goals to harmonize their investments with the goal of sustainability, seeking out companies and investment opportunities, doing things that need to be done to move the world in sustainable directions.

Fifth, institutions might be ranked on the basis of what their graduates do in the world. On average, what price will future generations pay for the manner in which graduates of particular institutions now live? How much do they consume over a lifetime? How much carbon dioxide do they contribute to the atmosphere? How many trees do they plant? How do they earn their keep? How many work through business, law, social work, education, agriculture, communications, research, and so forth to create the basis for a sustainable society? Are they part of the larger ecological enlightenment that must occur as the basis for any kind of sustainable society, or are they part of the rear guard of a vandal economy? Most colleges make serious efforts to discover who among their alumni have attained wealth. I know of no college that has surveyed its graduates to determine their cumulative environmental impacts.

This leads me finally to an observation and a modest suggestion. All educational institutions honor alumni in various ways, including the granting of honorary degrees mostly in direct proportion to wealth, power, fame, and gifts not yet received. None to my knowledge has ever revoked a degree for any cause whatsoever. Perhaps they should. If, for example, it were discovered that a graduate could not read, the embarrassment would be great and the institution's reputation would be greatly and deservedly damaged. No such shame as yet is attached to graduates who are merely ecologically illiterate and ignorant of how the planet works. There is, I think, only one reasonable course of action, the precedent for which is the practice of recalling defective automobiles at the manufacturer's expense. Likewise, defective minds should be "recalled" and offered an opportunity to return to the institution's tutelage to undergo remedial instruction. Alternatively, the institution that awarded the degree may wish to refund the tuition plus interest charges with its apology. It would, of course, remain liable for the damage done to the earth by the degree holder as a result of an ecologically defective education. In either case the nation, the institution, and the offender would be well served, and all would be greatly edified.

SOURCES

Smith, A. 1992. *Campus Ecology*. Los Angeles: Living Planet Press.

CHAPTER FOURTEEN

The Problem of Disciplines and the Discipline of Problems

W E EXPERIENCE nature mostly as sights, sounds, smells, touch, and tastes—as a medley of sensations that play upon us in complex ways. But we do not organize education the way we sense the world. If we did, we would have Departments of Sky, Landscape, Water, Wind, Sounds, Time, Seashores, Swamps, Rivers, Dirt, Trees, Animals, and perhaps one of Ecstasy. Instead we have organized education like mailbox pigeonholes, by disciplines that are abstractions organized for intellectual convenience. Hardly one scholar in ten could say why or when this came to be, but most would state with great conviction that it is quite necessary and irrevocable. The "information explosion" has further added to the impulse to divide knowledge into smaller and smaller disciplinary categories, and the end is not in sight.

There is, nonetheless, a good bit of grumping about academic specialization, intellectual narrowness, and pigeonhole thinking. But despite decades of talk about "interdisciplinary courses" or "transdisciplinary learning," there is a strong belief that such talk is just talk. Those thought to be sober, or at least judiciously dull, mostly presume that real scholarship means getting on with the advance of knowledge organized exclusively by disciplines and subdisciplines. It does not seem to matter that some knowledge may not contribute to an intelligible whole, that some of it is utterly trivial, that parts of it are contradictory, or that significant and life-enhancing things are omitted.

If this were all that happened as a consequence of the way we organize knowledge, the results would be merely unfortunate, but the truth is that the consequences are, in a deeper sense, tragic. The great ecological issues of our time have to do in one way or another with our failure to see things in their entirety. That failure occurs when minds are taught to think in boxes and not taught to transcend those boxes or to question overly much how they fit with other boxes. We educate lots of in-the-box thinkers who perform within their various specialties rather like a dog kept in the yard by an electronic barrier. And there is a connection between knowledge organized in boxes, minds that stay in those boxes, and degraded ecologies and global imbalances. The situation is tragic in that many suspect where all of this is leading but believe themselves powerless to alter it.

Our situation is tragic in another way. Often those who do comprehend our plight intellectually cannot feel it and hence are not moved to do much about it. This is not merely an intellectual failure to recognize our dependence on natural systems, which is fairly easy to come by. It is, rather, a deeper failure in the educational process to join intellect with affection and loyalty to the ecologies of particular places, which is to say a failure to bond minds and nature. It is no accident that this bonding happens far less often than we might hope. Professionalized and specialized knowledge is not about loyalty to places or to the earth, or even to our senses, but rather about loyalty to the abstractions of a discipline. The same can be said of the larger knowledge "industry," which was intended to make us rich and powerful by industrializing the world. This may help to explain why increasingly sophisticated analyses of our plight coincide with a paralysis of will and imagination to get at its roots.

And so we tinker. We add a course here and another requirement there and hold a symposium in some exotic place. Those who are bold enough tack on another outshed to the rambling curricular edifice of Babel and call it "environmental studies." If our crisis, however, is first and foremost a crisis of mind and perception, as I believe it to be, the time has come for a fundamental reconsideration of how we might encourage what Edith Cobb (1977) has called "an acute sensory response to the natural world" (p. 30). I offer two ideas.

First, I suggest that at all levels of learning K through PhD, some part of the curriculum be given to the study of natural systems roughly in the manner in which we experience them. The idea is hardly novel. In various ways it is the basis of programs offered by the National Outdoor Leadership School, Outward Bound, and can be found in courses in a few innovative institutions, such as Prescott College in Arizona, and in others offered by a few nonprofit institutes. It is also an old idea, going back at least as far as the belief that nature has something to teach us. The idea is simply that we take our senses seriously throughout education at all levels and that doing so requires immersion in particular components of the natural world—a river, a mountain, a farm, a wetland, a forest, a particular animal, a lake, an island—*before* students are introduced to more advanced levels of disciplinary knowledge.

For example, a course on a nearby river might require students to live on the river for a time, swim in it, canoe it, watch it in its various seasons, study its wildlife and aquatic animals, listen to it, and talk to people who live along it. A river becomes, as biologist Carl McDaniel (1993) phrased it, "a microcosm of the world" and a doorway to wider knowledge. Each student might research a particular aspect of the river, say, its folklore, social history, evolution, art, chemistry, ecology, literature, or the politics and law that govern its use. Collectively, a picture of the river might begin to emerge that would be more than the sum of the individual projects. I am not proposing just a weekend field trip but a longer period of time to allow the senses to soak in the experience as sights, sounds, tastes, smells, and feel until something like profound respect, or more, begins to take root.

What might such experiences do? First, they would remove the abstractness and secondhand learning that corrupts knowledge at its source. Natural objects have a concrete reality that the abstractions of textbooks and lectures do not and cannot have. Second, a course on a river or a forest or a farm might help make up the experience deficit now common among urban and suburban young people whose minds have been exposed overly long to shopping malls, video games, and television. Third, it would cultivate mindfulness by slowing the pace of learning to allow a deeper kind of knowing to occur. Fourth, it would give students stronger reasons to want to learn those things that require the knowledge of various disciplines. Fifth, it would teach the art of careful field observation and the study of place. Sixth, it would teach students that there are some things that cannot be known or said about a mountain, or a forest, or a river—things too subtle or too powerful to be caught in the net of science, language, and intellect. It would introduce students to the mysterious and unknowable before the mere unknowns of a particular discipline.

What I am proposing, more broadly, is rather like a courtship between mind and nature, or perhaps like an awakening. I believe that we should introduce students to the mysteries of specific places and things before giving them access to the power inherent in abstract knowledge. I am proposing that we aim to fit the values and loyalties of students to specific places before we equip them to change the world. I propose that we give students a stronger reason to want to know while making them more trustworthy in the use of knowledge. I am proposing that we make them accountable in small things before giving them the keys to the creation.

Among the precedents for the kind of experience I am proposing are Thoreau's *Walden*, Aldo Leopold's approach to natural history, Annie Dillard's (1974) sojourn at Tinker Creek, John Hanson Mitchell's (1984) study of 15,000 years on a square mile in Massachusetts, and William Least Heat Moon's study of Chase County, Kansas, what he calls "a deep map" (Moon, 1991). And there is the experience, if we are willing to acknowledge it, from indigenous cultures, many of which were extraordinarily adept at drawing mind and nature together.

I have a second and related suggestion for overcoming disciplinary narrowness and the aloofness that is all too often characteristic of academic institutions. Alan Mermann (1992) at the Yale School of Medicine described the problem in these words:

> Careful studies and accurate reports are done; papers are published in distinguished journals; but evidence of efforts to engage the problems on site is sadly lacking. . . . We have a long history of avoidance of unpleasant tasks requiring commitment and sacrifice. . . . We find it easier to use our minds and our resources for the solution of intellectual problems because we are then freed from the burdens of seeing our interdependence and our indebtedness as persons.

I believe this, in the main, to be true. Educational institutions and professionalized scholars do tend to seal themselves off from the unpleasant and less rewarding challenges around them. And when they do engage those challenges, they do so as "research," not as serious efforts to solve real problems.

In contrast, I propose that we engage young people and faculty together in the effort to solve real problems. I do not propose such efforts as "service" projects alone but as ways to integrate learning with service. Opportunities are all around us. Virtually all schools and institutions of higher education are located in places that are losing biological diversity and the means for right livelihood, rural and urban places alike that are polluted, overexploited, and increasingly derelict. What do we know that might restore such places? How might the effort to solve real problems be made a part of the conventional curriculum? How might the discipline of solving problems change the organization of education?

Problem solving requires broadening what we take to be our constituency to include communities in which educational institutions are located. It requires institutional flexibility and creativity, which in turn presuppose a commitment to make knowledge count for the long-term health of local communities and people. It requires overcoming the outmoded idea that learning occurs exclusively in classrooms, laboratories, and libraries. It requires acknowledgment of the possibility that learning sometimes occurs most thoroughly and vividly when diverse people possessing different kinds of knowledge pool what they know and join in a common effort to accomplish something that needs to be done. When they do, they discover ways to communicate that disciplinary education alone cannot produce. They quickly learn to distinguish what is important from what is not. And students and faculty alike discover that they are competent to change things that otherwise appear to be unchangeable.

We are not likely anytime soon to dispense with disciplinary knowledge, nor do I propose to do so. What I do propose is that we seek out ways to situate disciplinary knowledge within a more profound experience of the natural world while making it more relevant to the great quandaries of our age.

SOURCES

Cobb, E. 1977. *The Ecology of Imagination in Childhood*. New York: Columbia University Press.

Dillard, A. 1974. *Pilgrim at Tinker Creek*. New York: Harper's Magazine Press.

McDaniel, C. 1993. *One Mile of the Hudson*. Unpublished Manuscript.

Mermann, A. C. 1992. *Questions and Reflections*. New York: Pharos Books.

Mitchell, J. H. 1984. *Ceremonial Time*. New York: Anchor Press.

Moon, W. 1991. *Prairy-Erth*. Boston: Houghton Mifflin.

CHAPTER FIFTEEN

Professionalism and the Human Prospect

The mind can be permanently profaned by the habit of attending to trivial things, so that all our thought shall be tinged with triviality.
— HENRY DAVID THOREAU

I have always tried not to be a professional *scientist.*
— ERWIN CHARGAFF

THE TENURE system was originally created to protect the right of professors to speak freely without fear of reprisal. One might have expected great and radical things to emanate from the safely tenured. With some notable exceptions, however, this has not happened often. Derek Bok (1990), former President of Harvard University, for one, has lamented the results:

> Armed with the security of tenure and the time to study the world with care, professors would appear to have a unique opportunity to act as society's scouts to signal impending problems. . . . Yet rarely have members of the academy succeeded in discovering emerging issues and bringing them vividly to the attention of the public (p. 105).

Similarly, why have so few of the tenured joined the effort to preserve biological diversity and a habitable earth? Why are so few of the tenured willing to confront the large and portentous issues of human survival looming ahead?

I believe the reason is that the professorate professionalized itself, and professionalization has done what even the most flagrant college administrator would not dream of doing. The professionally induced fear of making a mistake or being thought to lack rigor has rendered much of the professorate toothless and confined to quibbles of great insignificance. One sure way for a young professor to risk being denied tenure by his peers is to practice what philosopher Mary Midgley (1989) called "the virtue of controversial courage" (p. 69), the very reason for which tenure was created. For the consummate professional scholar, the rule of thumb is that if it has no obvious and quick professional payoff, don't do it. A modern Charles Darwin, for example, would have gotten a large grant, flown to the Galapagos, returned to dash off a dozen articles to a dozen journals, hired a publicity agent to get on TV and in the *Science Times*, all with the intention of getting a better job somewhere else. The real Darwin delayed publication of *The Origin of Species* (1859) for 20 years while he thought it over. Those with something to profess, in short, are being replaced by professionals with something to sell.

Professional scholars tend to think of themselves as part of the established order, not as critics of it, let alone creators of something better. Knowledge has increasingly become mastery of allegedly neutral techniques. In the process, however, neutrality has gotten confused with objectivity. In the words of historian Robert Proctor (1991), "Neutrality refers to whether a science takes a stand; objectivity to whether a science merits certain claims to reliability" (p. 10). The ideal of the broadly informed, renaissance mind has given way to the far smaller idea of the specialist. Even in the humanities where one might still expect dangerous thoughts about the plight and potentials of humankind, one often finds instead silliness, shrillness, and obscurity.

Furthermore, professionalization has Balkanized the intellectual landscape, each fiefdom having its own professional association, trade journals, and specialized jargon. Professing allegiance to one principality or another, few "professionals" know enough of the whole terrain to be dangerous to the established order. Narrowness, "methodolatry," and careerism have rendered many unfit and unwilling to ask large and searching questions. Where intellectuals once addressed the public, they now talk mostly to each other about matters of little or no consequence for the larger society. To the same degree that it is obscure, jargon-laden, and trivial, professionalized knowledge has come as a great windfall to the comfortable, serving to divert attention from behavior that is egregious, criminal, or merely embarrassing. When did an issue of the *American Political Science Review* cause the comfortable in Congress to squirm? When did an issue of the *American Economic Review* ever cause the barons of Wall Street to tremble? When was the last time agribusiness felt itself threatened by the American Society of Agronomists? When was the last time the dispossessed felt befriended by an issue of the *American Sociological Review*? When did those now planning to reengineer the fabric of life on earth for profit feel their project threatened by any academic profession at all? And when did philosophy "cease to be 'the love of wisdom' and aspire to be a science" (Solomon, 1992, p. 19)?

The academy, in short, is a safer haven than it ought to be for the professionally comfortable, cool, and upwardly mobile. It is far less often than it should be a place for passionate and thoughtful critics. Professionalization has rendered knowledge safe for power, thereby making it more dangerous than ever to the larger human prospect. And what are these dangers?

First, there is the danger that professionals will clone themselves, making their students knowledge technicians instead of broadly thoughtful, liberally educated persons willing to roll up their sleeves and join in the great struggles ahead to preserve a habitable planet. Second, there is the danger that a great deal of important "nonprofessional" knowledge will be dismissed or ignored altogether. I am referring to the kind of vernacular knowledge that people have always needed to live well in a place, to create enduring communities, to understand themselves, to be of service to those around them, and to find meaning amidst the mysteries of life. Third, despite all of the recent talk about multiculturalism, the plain fact is that the academy has become an agent of global homogenization. The great questions of human existence are being reduced to those amenable to narrow professionalized scholarship. The danger here is that a global monoculture, driven by the logic of a worldwide market economy and confined to the language of professional discourse, will not and, I believe, cannot preserve a diverse biota for long. There is a fourth danger, which is simply that higher education will mostly opt out of the great ecological issues of the twenty-first century because it cannot summon enough vision and courage to do otherwise.

What can be done? The tenure system continues to provide a defense against capricious administrators. It was never intended, however, to protect against the more subtle and powerful censorship of academic mandarins whose function it is to protect disciplinary boundaries. For this I offer, accordingly, two ideas to curtail professionalism run amuck. The first is to suggest that all candidates for tenure appear before an institution-wide forum to answer questions such as the following:

- Where does your field of knowledge fit in the larger landscape of learning?
- Why is your particular expertise important? For what and for whom is it important?
- What are its wider ecological implications and how do these affect the long-term human prospect?
- Explain the ethical, social, and political implications of your scholarship.

The benefits of such a forum are clear. It would provide a great incentive for the candidate to think beyond the confines of his or her particular discipline. It would also provide considerable incentive to communicate plainly in commonly understood language, which would be something of a novelty. It would help to establish a balance between those with a sense of the larger issues of the time and those who are content to be narrowly specialized professionals. It would smoke out certain kinds of intellectual deficiencies, such as ethical flaccidity or gross ignorance of the relation of ecological realities to one's scholarship. A tenure forum might even exert a moderately beneficial effect on the inquisitors, causing them to ponder the larger architecture and purposes of knowledge as well.

My second suggestion similarly aims to legitimize and encourage disciplinary boundary crossings and thereby weaken the hold of a narrow professionalism. Educators might learn something from the way many prosperous business organizations have developed flexible teams that regularly reshuffle organizational responsibilities to achieve results not otherwise possible in a rigid or hierarchical structure. Colleges and universities might similarly pioneer new and daring ways to reorganize knowledge to attack problems of the day, but without the heavy baggage of disciplines and departments. I propose that all educational institutions create and amply fund regular forums, programs, courses, and projects that offer participants the opportunity to suspend their status as disciplinary specialists. University of California at Los Angeles sociologist Jeffrey Alexander and ten of his colleagues, for example, offer a course called "L.A. in Transition" (Alexander, 1993, p. B3). The course aims "to make Los Angeles into a guinea pig for broad study of contemporary racial, ethnic, and economic problems." In such settings scholars would have to step outside their particular disciplines in order to participate in a common dialogue, solve a problem, or cooperate in the creation of something new. Most colleges and universities have interdisciplinary programs that provide a setting for such endeavors, but participation is seldom seriously rewarded, acknowledged, or funded. And such programs are usually the first to be dropped when financial cuts are made. The remedy is straightforward: unequivocal institutional support and encouragement for boldness, breadth, and disciplinary boundary crossings.

For those willing to do it, the task of mastering a particular field in depth while acquiring a broad and contextual knowledge demands time, patience, intellectual skill, and great commitment. It demands scholars who pay attention to large issues and who have loyalties to things bigger than the profession. These people deserve to be protected from both capricious administrators and from what Page Smith (1990) has called "academic fundamentalists" (p. 5). The world has always needed a dangerous professorate and needs one now more than ever before. It needs a professorate with ideas that are dangerous to greed, shortsightedness, indulgence, exploitation, apathy, high-tech pedantry, and narrowness.

SOURCES

Alexander, J. C. 1993, December 1. The Irrational Disciplinarity of Undergraduate Education. *The Chronicle of Higher Education*, p. B3.

Bok, D. 1990. *Universities and the Future of America*. Durham: Duke University Press.

Darwin, C. 1986. *Origin of Species*. New York: NAL/Dutton. (Original work published 1859.)

Midgely, M. 1989. *Wisdom, Information, and Wonder*. London: Routledge.

Proctor, R. 1991. *Value Free Science*. Cambridge: Harvard University Press.

Smith, P. 1990. *Killing the Spirit*. New York: Viking.

Solomon, R. 1992. Beyond Reason: The Importance of Emotion in Philosophy. In J. Ogilvy, ed., *Revisioning Philosophy*. Albany: State University of New York Press.

CHAPTER SIXTEEN

Designing Minds

A S *HOMO SAPIENS*'S entry in any intergalactic design competition, industrial civilization would be tossed out at the qualifying round. It doesn't fit. It won't last. The scale is wrong. And even its apologists admit that it is not very pretty. The design failures of industrially/ technologically driven societies are manifest in the loss of diversity of all kinds, destabilization of the earth's biogeochemical cycles, pollution, soil erosion, ugliness, poverty, injustice, social decay, and economic instability.

Industrial civilization, of course, was not designed at all; it simply happened. Those who made it happen were mostly singleminded men and women innocent of any knowledge of what can be called the "ecological design arts," by which I mean the set of perceptual and analytical abilities, ecological wisdom, and practical wherewithal essential to making things that "fit" in a world of trees, microbes, rivers, animals, bugs, and small children. In other words, ecological design is the careful meshing of human purposes with the larger patterns and flows of the natural world and the study of those patterns and flows to inform human purposes.

Ecological design competence means maximizing resource and energy efficiency, taking advantage of the free services of nature, recycling wastes, making ecologically smarter things, and educating ecologically smarter people. It means incorporating intelligence about how nature works, what David Wann (1990) called "biologic," into the way we think, design, build, and live. Design applies to the making of nearly everything that directly or indirectly requires energy and materials or governs their use, including farms, houses, communities, neighborhoods, cities, transportation systems, technologies, economies, and energy policies. When human artifacts and systems are well designed, they are in harmony with the larger patterns in which they are embedded. When poorly designed, they undermine those larger patterns, creating pollution, higher costs, and social stress in the name of a spurious and short-run economizing. Bad design is not simply an engineering problem, although better engineering would often help. Its roots go deeper.

Good design, begins as Wendell Berry (1987) stated, by asking, "What is here? What will nature permit us to do here? What will nature help us to do here?" (p. 146). Good design everywhere has certain common characteristics including the following:

- right scale,
- simplicity,
- efficient use of resources,
- a close fit between means and ends,
- durability,
- redundance, and
- resilience.

Good design also solves more than one problem at a time. They are often place specific or, in John Todd's words, "elegant solutions predicated on the uniqueness of place." Good design promotes

- human competence instead of addiction and dependence,
- efficient and frugal use of resources,
- sound regional economies, and
- social resilience.

Where good design becomes part of the social fabric at all levels, unanticipated positive side effects (synergies) multiply. When people fail to design carefully, lovingly, and competently, unwanted side effects and disasters multiply.

By the evidence of pollution, violence, social decay, and waste all around us, we have designed things badly. Why? There are, I think, three primary reasons. The first is that while energy and land were cheap and the world relatively "empty," we simply did not have to master the discipline of good design. We developed extensive rather than intensive economies. Accordingly, cities sprawled, wastes were dumped into rivers or landfills, farmers wore out one farm and moved on to another, houses and automobiles got bigger and less efficient, and whole forests were converted into junk mail and Kleenex. Meanwhile, the know how necessary to a frugal, well-designed, intensive economy declined and words like *realistic* or *convenience* became synonymous with habits of waste.

Second, design intelligence fails when greed, narrow self-interest, and individualism take over. Good design is a community process requiring people who know and value the positive things that bring them together and hold them together. Old-order Amish farmers, for example, refuse to buy combines not because they would not make things easier or more profitable but because they would undermine community by depriving people of the opportunity to help their neighbors. This is pound wise and penny foolish the way intelligent design should be. In contrast, American cities with their extremes of poverty and opulence are products of people who believe that they have little in common with other people. Suspicion, greed, and fear undermine good community and good design alike.

Third, poor design results from poorly equipped minds. Good design can only be done by people who understand harmony, patterns, and systems. Good design requires a breadth of view that leads people to ask how human artifacts and purposes "fit" within the immediate locality and within the region. Industrial cleverness, however, is mostly evident in the minutiae of things, not in their totality or in their overall harmony. Moreover, good design uses nature as a standard and so requires ecological intelligence, by which I mean a broad and intimate familiarity with how nature works. For all of the recent interest in environment and ecology, this kind of knowledge, which is a product of both local experience and stable culture, is fast disappearing.

As an example of this kind of knowledge, George Sturt (1984), one of the last wheelwrights in England, described in *The Wheelwright's Shop* what he called "the age-long effort of Englishmen to fit themselves close and ever closer into England" (p. 66). Sturt built wagons crafted to fit the buyer's particular habits, needs, and topography. To do so, he needed to know a great deal about how his customers used a wagon, whether they drove fast or slow, whether their land was rocky or wet, and what they hauled. As a result,

> we got curiously intimate with the peculiar needs of the neighborhood. In farm-waggon or dung-cart, barley-roller, plough, water barrel, or what not, the dimensions we chose, the curves we followed, were imposed upon us the nature of the soil in this or that farm, the gradient of this or that hill, the temper of this or that customer or his choice perhaps in horseflesh. (p. 18)

Furthermore, the wheelwright needed to know what kinds of trees gave particular parts extra strength, or flexibility, or weight, where these trees grew, and when they were ready to harvest. And finally he needed to know the traditions and skills unique to his craft that were passed down as folk knowledge:

> What we had to do was to live up to the local wisdom of our kind; to follow the customs, and work to the measurements, which had been tested and corrected long before our time in every village shop all across the country. (p. 19)

The kind of mind that could design and build a good wagon depended a great deal on time-tested knowledge and intimate familiarity with place. The results were wagons that fit particular people and a particular landscape.

A contemporary example of ecological design can be found in John Todd's "living machines," which are carefully orchestrated ensembles of plants, aquatic animals, technology, solar energy, and high-tech materials to purify wastewater, but without the expense, energy use, and chemical hazards of conventional sewage treatment technology. According to Todd (1991),

> People accustomed to seeing mechanical moving parts, to experiencing the noise or exhaust of internal combustion engines or the silent geometry of electronic devices, often have difficulty imagining living machines. Complex life forms, housed within strange light-receptive structures, are at once familiar and bizarre. They are both garden and machine. They are alive yet framed and contained in vessels built of novel materials. . . . Living machines bring people and nature together in a fundamentally radical and transformative way. (pp. 335–343)

Todd has created several working examples of living machines, each resembling a greenhouse filled with exotic plants and aquatic animals. Wastewater enters at one end; purified water leaves at the other. In between, the work of sequestering heavy metals in plant tissues, detoxifying toxics, and removing nutrients has been done by biological systems driven by sunlight. A decade earlier he designed and built structures that similarly used aquatic systems to process waste, grow food, and store heat. Living machines and biologic imply changes in the way we process wastewater, grow food, and build houses and in the ways we integrate these and other functions into systems patterned after natural processes to do what industrial technology can only do expensively and destructively.

Ecological design also applies to the design of governments and public policies. Governmental planning and regulation require large and often ineffective or counterproductive bureaucracies. Design, in contrast, means

> the attempt to produce the outcome by establishing the criteria to govern the operations of the process so that the desired result will occur more or less automatically without further human intervention. (Ophuls, 1977, pp. 228–229)

In other words, well-designed policies and laws get the macro things right like prices, taxes, and incentives while preserving a high degree of micro freedom in how people and institutions respond. Design focuses on the structure of problems as opposed to their coefficients. For example, the Clean Air Act of 1970 required car manufacturers to install catalytic converters to remove air pollutants. Twenty-two years later emissions per vehicle are down substantially, but with more cars on the road, air quality is about the same. A design approach to transportation would lead us to think more about creating access between housing, schools, jobs, and recreation that eliminate the need to move lots of people and materials over long distances. A design approach would have led us to reduce dependence on automobiles by building better public transit systems, restoring railroads, and creating bike trails and walkways. A design approach would also lead us to rethink the use of urban land and to reintegrate agriculture and wilderness into urban areas.

❖ The Liberal and the Ecological Design Arts ❖

Ecological design requires the ability to comprehend patterns that connect, which means getting beyond the boxes we call disciplines to see things in their ecological context. It requires, in other words, a liberal education, but nearly everywhere the liberal arts have tended to become more specialized and narrow. Design competence requires the integration of first-hand experience and practical competence with theoretical knowledge, but the liberal arts have become more abstract, fragmented, and remote from lived reality. Design competence requires us to be students of the natural world, but the study of nature is being displaced by the effort to engineer nature to fit the economy instead of the other way around. Finally, design competence requires the ability to inquire deeply into the purposes and consequences of things to know what is worth doing and what should not be done at all. But the ethical foundations of education have been diluted by the belief that values are relative. All of this is to say that from an ecological perspective the "liberal arts" have not been liberal enough. I think this is evident in four respects.

First, the liberal arts have not been liberal enough in their response to the rapid decline in the habitability of the earth. Global and national policy change are necessary but insufficient to reverse downward trends in the earth's vital signs. It is also essential that we educate a citizen constituency that supports change and is competent to do the local work of rebuilding households, farms, institutions, communities, corporations, and economies that (1) do not emit carbon dioxide or other heat-trapping gases; (2) do not reduce biological diversity; (3) use energy, materials, and water with high efficiency; and (4) recycle wastes. In other words, a constituency that is capable of building economies that can be sustained without further reducing the earth's potential to sustain life. At a minimum this will require a modification of the skills, aptitudes, abilities, and curriculum by which we learned how to industrialize the earth.

Second, the liberal arts have come to mean an education largely divorced from practical competence. Inclusion of the ecological design arts in the liberal arts means bringing practical experience back into the curriculum in carefully conceived ways. The reasons, in Alfred North Whitehead's (1967) words, are straightforward: "First-hand knowledge is the ultimate basis of intellectual life. . . . the secondhandedness of the learned world is the secret of its mediocrity" (p. 51). In contrast to the distinction that John Henry Newman once drew between desirable and useful knowledge (Newman, 1982, pp. 84–88), Whitehead argued that there is a "reciprocal influence between brain activity and material creative activity" essential for good thinking. In other words, good thinking and practical experience are mutually necessary. Accordingly, he thought, "The disuse of hand-craft is a contributory cause to the brain-lethargy of aristocracies." J. Glenn Gray (1984) has argued similarly that the exclusion of manual skills from the liberal arts is dangerous "because it first divorces us from our own dispositions at the level where intellect and emotions fuse" (p. 85). Purely analytical and abstract thinking "separates us from our natural and human environment" (p. 85). Genuinely liberal education, in contrast, cultivates the full person, including manual competence and feeling as well as intellect.

Third, the liberal arts have come to include any number of fields, sub-fields, issues, and problems excepting those that are closest at hand in the local community. Inclusion of the ecological design arts suggests a symbiotic relation between learning and locality. Here, too, the reasons are part of an older tradition going back to John Dewey. In 1899 John Dewey wrote that "the school has been so set apart, so isolated from the ordinary conditions and motives of life" that children cannot "get experience—the mother of all discipline" (Dewey, 1990, p. 17). His solution required integrating opportunities for students "to make, to do, to create, to produce" and ending the separation of theory and practice. Dewey proposed that the immediate vicinity of the school be a focus of education, including the study of food, clothing, shelter, and nature. Through the study of these things, students might learn "the measure of the beauty and order about him, and respect for real achievement." Gray (1984) has likewise argued that liberal education is "least dependent on formal instruction. It can be pursued in the kitchen, the workshop, on the ranch or farm" (p. 81). It can also be pursued through the study of energy, water, materials, food, and waste flows on the campus.

How can competence in the ecological design arts be taught within the conventional curriculum? There are at least two broad possibilities. The best, but most difficult, approach is to make over entire institutions so that their operations and resource flows (food, energy, water, materials, waste, and investments) become a laboratory for the study of ecological design. There is a strong case for doing this for economic as well as pedagogic reasons (Orr, 1990). A second possibility follows the suggestion of Herman Daly and John Cobb to establish separate centers or institutes within colleges and universities with the mission of fostering ecological design intelligence (Daly and Cobb, 1989, pp. 357–360). Ecological design arts centers would aim to (1) develop a series of ecological design projects that involve students, faculty, and staff; (2) study institutional resource flows; (3) develop curriculum; and (4) carry out studies on environmental trends throughout the region. Ecological design projects could include, for example,

- design of a building with no outside energy sources, using locally available environmentally benign materials, that recycles all waste generated by occupants;

- development of a bioregional directory of building materials;
- inventory campus resource flows;
- restoration of a degraded ecosystem on or near the campus;
- design of a low-input, sustainable farm system;
- economic survey of resource and dollar flows in the regional economy; and
- design of solar aquatic wastewater systems for campus effluents.

The list could be easily extended, but the point is clear. The functions of ecological design institutes are (1) to equip young people with a basic understanding of systems and to develop habits of mind that seek out "patterns that connect" human and natural systems; (2) to teach young people the analytical skills necessary for thinking accurately about cause and effect; (3) to give students the practical competence necessary to solve local problems; and (4) to teach young people the habit of rolling up their sleeves and getting down to work.

SOURCES

Berry, W. 1987. *Home Economics*. San Francisco: North Point Press.

Daly, H., and Cobb, J. 1989. *For the Common Good*. Boston: Beacon Press.

Dewey, J. 1990. *The School and Society*. Chicago: University of Chicago Press. (Original work published 1899.)

Gray, J. G. 1984. *Rethinking American Education*. Middletown: Wesleyan University Press.

Newman, J. H. 1982. *The Idea of A University*. Notre Dame, IN: Notre Dame University Press.

Ophuls, W. 1977. *Ecology and the Politics of Scarcity*. San Francisco: W. H. Freeman.

Orr, D. 1990. The Campus, the Liberal Arts, and the Biosphere. *Harvard Educational Review* 60, 2, pp. 205–206.

Sturt, G., 1984. *The Wheelwright's Shop*. New York: Cambridge University Press. (Original work published 1923.)

Todd, J. 1991. Ecological Engineering, Living Machines and the Visionary Landscape. In C. Etnier and B. Guterstam, eds., *Ecological Engineering for Wastewater Treatment*. Gothenburg, Sweden: Boksgaden.

Wann, W. 1990. *Biologic*. Boulder: Johnson Publishing Co.

Whitehead, A. N. 1967. *The Aims of Education*. New York: Free Press.

CHAPTER SEVENTEEN

Architecture as Pedagogy

T IS paradoxical that buildings on college and university campuses, places of intellect, characteristically show so little thought, imagination, sense of place, ecological awareness, and relation to any larger pedagogical intent. The typical academic building seems to have the architectural elegance and performance standards common to shopping malls, motels, and drive-through funeral parlors: places where, one might infer, considerations of "throughput" are uppermost in the minds of designers. How has this come to be? Some believe it is the result of a conspiracy of sorts between a wealthy donor wishing to make an end run around mortality by having his or her name on a building; a college president wishing to enhance a reputation for getting things done; an architect seeking a professional reputation by designing showy buildings that do not work very well; and a financial officer whose job it is to economize on beauty, humanity, and common sense in the name of fiscal integrity. Personally, I do not believe that the design of academic buildings is the result of a conspiracy at all. Most academic administrators and trustees are fully capable of doing all of this on their own in broad daylight. They have not conspired, because they did not need to—faculty and students have been effectively excluded from the process whereby ambition tempered by dullness and tortured by utility is rendered into architectural form.

The problem is not just that many academic buildings are unsightly, do not work very well, or do not fit their place or region. The deeper problem is that academic buildings are not neutral, aseptic factors in the learning process. We have assumed, wrongly I think, that learning takes place in buildings, but that none occurs as a result of how they are designed or by whom, how they are constructed and from what materials, how they fit their location, and how—and how well—they operate. My point is that academic architecture is a kind of crystallized pedagogy and that buildings have their own hidden curriculum that teaches as effectively as any course taught in them. What lessons are taught by the way we design, build, and operate academic buildings?

The first lesson is that architecture is the prerogative of power and not that of those who teach or learn. Implicit in this view is the assumption that architecture does not influence the flow of ideas, the quality of learning, and the human relationships in which learning is embedded. Therefore, faculty and students are rarely consulted on whether or what to build or where. From this, they learn that power can impose what it wishes on the academic landscape without having to explain much.

The second lesson is that architecture and building design are merely technical and are thus best left to people with technical competence. It follows that ethical, ecological, or aesthetic aspects of building do not matter nearly as much as technique and technology. In deference to expertise, then, we learn passivity toward the "built environment." This may explain our subsequent failure to protest the spread of ugliness and banality across the landscape as well as our apparent obliviousness to how these cheapen our lives and diminish our prospects.

From the design and materials used in construction, a third lesson is learned: The environmental and energy costs of building do not matter much. Academic buildings are seldom designed to maximize solar gain or energy efficiency, to minimize unpriced environmental costs of materials, or to utilize local materials. Thus, we learn the carelessness that accompanies waste and inefficiency, as well as callousness to the degradation of other places from which materials and energy originate.

Fourth, a "successful" building is one that quietly serves the educational process but requires no mindfulness of those who use it. From this we learn passivity and disengagement from our surroundings and the irresponsibility invited by never having to know how things work, or why, or what alternatives there might have been. The same building in which sophisticated theories are propounded unobtrusively teaches its occupants that it is OK to be oblivious to the most basic aspects of life support.

Fifth, the process teaches us about the limits of imagination. It is assumed, without anyone ever saying as much, that intellect can be nurtured in sterile places largely devoid of imagination. Therefore, creativity in academic architecture is mostly confined to facades replete with lots of glassy flourishes of form disengaged from any purpose beyond that of impressing the easily impressible. The use of imagination mostly stops short of the places where learning is supposed to happen, the design of which is still the cubical classroom or the lecture hall (a cavernous space with audiovisual equipment), both of which reached near state of the art sometime before the "Dark Ages." Such spaces do little to lift the spirits, stir the imagination, fuel the intellect, or remind us that we are citizens of ecological communities.

We have not thought of academic buildings as pedagogical, but they are. We have not exercised much imagination about the design of academic buildings, and it shows in a manifest decline in our capacity to envision alternatives to the urban and suburban excrescence oozing all around us. We have assumed that people who know little about learning and pedagogy were competent to design places where learning is supposed to occur. They are not, not alone anyway. What alternatives do we have?

Let us begin by asking what might be learned from the design, construction, and operation of the places where formal education takes place. First, the process of design and construction is an opportunity for a community to deliberate over the ideas and ideals it wishes to express and how these are rendered into architectural form. What do we want our buildings to say about us? What will they say about our ecological prospects? To what large issues and causes do they direct our attention? What problems do they resolve? What kind of human relationships do they encourage? These are not technical details but first and foremost issues of common concern that should be decided by the entire campus community. When they are addressed as such, the design of buildings fosters civic competence and extends the idea of citizenship.

Second, the architectural process is an opportunity to learn something about the relationship between ecology and economics. For example, how much energy will a building consume over its lifetime? How much of what kinds of materials will be required for its upkeep? What unpriced costs do construction materials impose on the environment? Are they toxic to manufacture, install, or, later, to dispose? How are these costs paid? What is the total energy embodied in materials used in the structure? Is it possible to design buildings that repay those costs by being net energy exporters? If not, are there other ways to balance ecological accounts? Can buildings and the surrounding landscape be designed to generate a positive cash flow?

These questions cannot be answered without engaging issues of ethics. How are building materials extracted, processed, manufactured, and transported? What ecological and human costs do various materials impose where and on whom? What in our ethical theories justifies the use of materials that degrade ecosystems, jeopardize other species, or risk human lives and health? Where those costs are deemed unavoidable to accomplish a larger good, how can we balance ethical accounts?

Fourth, within the design, construction, and operation of buildings is a curriculum in applied ecology. Buildings can be designed to recycle organic wastes through miniature ecosystems that can be studied and maintained by the users. Buildings can be designed to heat and cool themselves using solar energy and natural air flows. They can be designed to inform occupants of energy and resource use. They can be landscaped to provide shade, break winter winds, propagate rare plants, provide habitat for animals, and restore bits of vanished ecosystems. Buildings and landscapes, in other words, can extend our ecological imagination.

Fifth, they can also extend our ecological competence. The design and operation of buildings is an opportunity to teach students the basics of architecture, landscape architecture, ecological engineering for cleaning wastewater, aquaculture, gardening, and solar engineering. Buildings that invite participation can help students acquire knowledge, discipline, and useful skills that cannot be acquired other than by doing.

Finally, good design can extend our imagination about the psychology of learning. The typical classroom empties quickly when not required to be used. Why? The answer is unavoidable: It is most often an uninteresting and unpleasant place, designed to be functional and nothing more. And the same features that make it unpleasant make it an inadequate place in which to learn. What makes a place a good educational environment? How might the typical classroom be altered to encourage ecological awareness, creativity, responsiveness, and civility? How might materials, light, sounds, water, spatial configuration, openness, scenery, colors, textures, plants, and animals be combined to enhance the range and depth of learning? My hunch is that good learning places are places that feel good to us: human-scaled places that combine nature, interesting architecture, materials, natural lighting, and "white sounds" (e.g., running water) in interesting ways that resonate with our innate affinity for life.

My point is that the design, the construction, and the operation of academic buildings can be a liberal education in a microcosm that includes virtually every discipline in the catalog. The act of building is an opportunity to stretch the educational experience across disciplinary boundaries and across those dividing the realm of thought from that of application. It is an opportunity to work collectively on projects with practical import and to teach the art of "good work." It is also an opportunity to lower lifecycle costs of buildings and to reduce a large amount of unnecessary damage to the natural world incurred by careless design.

CHAPTER EIGHTEEN

Agriculture and the Liberal Arts

It is incumbent on us to take special pains . . . that all the people,
or as many of them as possible, shall have contact with the earth
and that the earth's righteousness shall be abundantly taught.
— LIBERTY HYDE BAILEY

NTIL quite recently much of what people knew about the natural
world they learned from the experience of growing up on a farm or
by periodic visits to nearby farms. For all of their flaws, farms were
schools of a sort in natural history, ecology, soils, seasons, wildlife, animal
husbandry, and land use. The decline of ecologically diverse farms and
the experience of the natural world that they fostered explains in large
part, I think, the increasing gap between the broad support for environmental causes evident in public opinion polls and a growing ignorance of
how ecosystems work and how private consumption and economic
growth destroy the environment. In other words, the sharp decline in the
number of farms and the shift toward industrial farming has had serious
consequences for our collective ecological intelligence.

To be sure, the experience of farm life varied greatly with the quality
of the farm and differences in individual perceptivity, intelligence, and
skill. Moreover, in the absence of vital rural communities, farm life was
sometimes tedious, narrow, and parochial. On balance, however, I believe
that it was mostly otherwise. But in either case farms did what no other
institution has ever done as well. They taught directly, and sometimes
painfully, the relationship between our daily bread and soil, rainfall, animals, biological diversity, and natural cycles, which is to say land stew-

ardship. They also taught the importance of the human qualities of husbandry, patience, hard work, self-reliance, practical skill, and thrift. However imperfectly, farms served as a reality check on human possibilities in nature that urban societies presently lack.

The decline of small family farms and rural communities might still be justified as the necessary price for efficiency. However, it is evident that unfair taxation and lavish subsidies for large scale had more to do with the demise of small farms than did their alleged failure to produce or to make a profit (Strange, 1988). On behalf of a short-term, devil-take-the-hindmost economics we destroyed farms and rural communities, which have been the historical ballast for stable societies. We have good reason to agree with Aldo Leopold (1991) that social "stability seems to vary inversely to the mental distance from fields and woods" (p. 286).

At its best, traditional farming and rural life were, in Jacquetta Hawkes's (1951) perception, "a creative, a patient and increasingly skilful love-making that had persuaded the land to flourish" (p. 202). In contrast, the industrialization of agriculture was "an upsurge of instinctive forces comparable to the barbarian invasions . . . designed to satisfy man's vanity, his greed and possessiveness, his wish for domination" (p. 203). As agriculture became more industrialized, the number of farms declined, and with them, rural communities. Remaining farms became larger, ecologically less diverse, more expensive to operate, and more vulnerable to economic and ecological forces beyond the control of farmers. They also tended to become less interesting and less instructive places, hence, the decline of land intelligence now evident throughout predominantly urban societies.

Farms were not only sources of instruction about the realities of the natural world; in many places they also served to protect biological diversity. Gary Nabhan's (1982) studies of the Papago and Gene Wilken's (1987) studies of peasant agriculture in Mexico and Central America show great ecological intelligence, carefully and artfully woven into rural landscapes over a millennium or more. Traditional peasant farms were repositories of genetic diversity, often growing dozens or hundreds of varieties that are now disappearing in favor of a small number of hybrids purchased from multinational corporations.

In many parts of the third world, the imposition of high-yield agriculture helped to break apart the intimate relationship between cultures and ecosystems that had coevolved over long periods of time. As Angus Wright (1990) has shown in a brilliant study, *The Death of Ramon Gon-*

zalez, these systems and the biological diversity they preserved have been largely destroyed as part of a development strategy designed to "milk the soil and other natural resources of the nation and its poorest laborers by squeezing money out of agriculture" to pay for industrialization (p. 227). In Mexico the result was, according to Wright, an effort to "concentrate earnings and investment in a few regions of agribusiness growth at the expense of the neglect, ruin, and abandonment of other regions" (p. 237).

What Wright called "the modern agricultural dilemma" is simply the following:

> The highly localized adaptations needed for ecologically healthy agriculture and healthy, stable rural communities are often in con-flict with the apparent requirements of rapidly industrializing nations and an expanding international economy. (p. 245)

From the perspective of a narrow science and economics, this dilemma is hard to see, or more precisely it is difficult for some to want to see because it implies failure of great consequence. From a wider perspective that includes a thorough knowledge of ecology and anthropology, the dilemma was in fact foreseen early on by people like Carl O. Sauer, Paul Sears, and Aldo Leopold (Wright, 1990, pp. 247, 285).

The decline of the family farm in the United States and the destruction of traditional farming practices in what is called the underdeveloped world is the product of many forces including the separation of the study of agriculture from its community, cultural, and ecological context. The modern agricultural dilemma described by Wright began when agricul-tural sciences were isolated in research institutions and from there evolved into technical disciplines whose purpose was to do one thing: increase production. Consequently they were not rooted in any coherent and sustainable social, philosophical, political, and ecological context, which would have meant doing many things simultaneously. In this set-ting a great many assumptions about nature, technology, farming, rural life, and the consequences of applying industrial techniques to complex biological and cultural systems went unchallenged.

It might have been very different had agriculture evolved instead within liberal arts colleges. Instead of becoming a series of disjointed tech-nical specialties, agriculture might have come to be regarded, and rightly so, as a liberal art with technical aspects. In the context of liberal arts colleges, agriculturalists might have learned to see farming not as a pro-duction problem to be fixed, but as a more complex activity, at once cultural, ethical, ecological, and political. This is not a new idea. Leopold (1991) proposed that it is time to "swap ends . . . to curtail sharply the output of professionals . . . [in order to] tell the whole campus and . . . the whole community what . . . conservation is all about" (p. 301). By the same logic, we have too many agricultural professionals and not nearly enough people who understand farming in its wider social and ecological context. The goal of liberal education, as Leopold described it, was "not merely a dilute dosage of technical education" but rather "to teach the student to see the land, to understand what he sees, and enjoy what he understands" (p. 302).

If agriculture might have evolved differently in a liberal arts setting, I think the inclusion of agriculture would have helped liberal arts colleges avoid the debilitating separation of abstract intellect and practical intelligence. Instead, we have developed a version of the liberal arts in which it is assumed, without anyone ever quite saying as much, that learning is an indoor sport taking place exclusively in classrooms, libraries, laboratories, and computer labs and that practical competence is to be avoided at all cost.

This leads me to propose that agriculture should be included as a part of a complete liberal arts education, first because it offers an important kind of experience no longer available to many young people from predominantly urban areas. Student responsibility for farm operations would teach the values of discipline, physical stamina, frugality, self-reliance, practical competence, hard work, cooperation, and ecological competence. Second, college farms properly used would be interdisciplinary laboratories for the study of sustainable agriculture, ecology, botany, zoology, animal husbandry, entomology, soil science, ornithology, landscape design, land restoration, mechanics, solar technology, business operations, philosophy, and rural sociology. Third, college farms could become catalysts in a larger effort to revitalize rural areas in surrounding areas. Fourth, college farms could be used to preserve biological diversity jeopardized by development. Fifth, college farms could be a part of a global effort to reduce carbon emissions involved in the long-distance transport of food by sequestering carbon through agroforestry and tree cropping. Sixth, college farms could close waste loops by composting all campus organic wastes and incorporating these as soil admendments. Finally, by participating in the design and operation of college farms, students could learn that our problems are not beyond intelligent solution; that solutions are close by; and that institutions that often seem to be inflexible, unimaginative, and remote from the effort to build a sustainable society can be otherwise.

SOURCES

Bailey, L. H. 1980. *The Holy Earth*. Ithaca: New York State College of Agriculture. (Original work published 1915.)

Hawkes, J. 1951. *A Land*. New York: Random House.

Leopold, A. 1991. *The River of the Mother of God and Other Essays by Aldo Leopold*. S. Flader and J. B. Callicott, eds. Madison: University of Wisconsin Press. (Original work published 1941.)

Nabhan, G. 1982. *The Desert Smells Like Rain*. San Francisco: North Point Press.

Strange, M. 1988. *Family Farming*. Lincoln: University of Nebraska Press.

Wilken, G. 1987. *Good Farmers: Traditional Agriculture Resource Management in Mexico and Central America*. Berkeley: University of California Press.

Wright, A. 1990. *The Death of Ramon Gonzalez: The Modern Agricultural Dilemma*. Austin: University of Texas Press.

CHAPTER NINETEEN

Educating a Constituency
for the Long Haul

I N *EARTH IN THE BALANCE*, Al Gore (1992), the U.S. Vice President, proposed making "the rescue of the environment the central organizing principle for civilization" (p. 269). Gore is, in effect, calling for a global constituency for the long haul, one oriented to the health of the planet, with a decent regard for the rights and interests of future generations and a degree of self-denial quite unusual in an upstart species that still mostly believes itself to be the master of all that it surveys and deserving of all that it can take.

That such a constituency is essential to our prospects is clear enough, but it may never come to exist to the extent necessary to rescue the earth or, more to the point, to rescue the human prospect on the earth. It certainly will not be created easily and quickly. Aside from the possibility that evolution has made us much more adept at dealing with visible threats like marauding armies than with invisible ones measured in parts per billion or possibilities of future catastrophe (Ornstein and Ehrlich, 1989), there are serious obstacles to the creation of any effective constituency for the long haul.

First, every parent knows that it is much easier for a two-year-old to make messes than to clean them up. The same is true in the realm of intellect and public policy. It is easier for demagogues to cast doubt, confuse, obfuscate, and muddy the water than it is to clarify great and complex issues. And in both cases it takes far longer to clean up the mess than it did to make it in the first place.

A problem like this is now evident in recent efforts to render science subservient to politics, ideology, and at times even fantasy. The politicization of science has become a growth industry. Rush Limbaugh, for example, a radio and television talk show host of enormously modest modesty, believes that ozone depletion is a scam foisted on a gullible public by scientists wanting more research money. I do not think that scientists are always above fooling the public. But in this case it is worth asking how Mr. Limbaugh, who may be similarly inclined, has arrived at his view. Reportedly (Taubes, 1993), he learned it from Dixie Lee Ray's (1992) book *Trashing the Planet*, the "most footnoted, documented book" Mr. Limbaugh says he has ever read, and we have reason to take him at his word. And how did Dixie Lee Ray learn such things? Well, she learned it in part from the writings of Rogelio Maduro, who holds a BS degree in geology and writes for a magazine published by supporters of Lyndon LaRouche, now living at public expense for conspiring to avoid paying his taxes. Mr. Maduro, in turn, derived his views from research that is now widely conceded to be wrong.

The issues surrounding ozone depletion are complex and do allow for honest disagreement. But this dispute is something else: scientific error compounded by the ideology of buccaneer capitalism, magnified by an enlarged ego, and multiplied by the power of modern communications. Trusting Mr. Limbaugh's facts and opinions and unwilling to investigate further, millions of his listeners now believe that we need not worry about the effects of ozone depletion, climate change, or any other environmental issue for that matter. Mr. Limbaugh is entitled to his many opinions. But alert listeners will note that he is not as quick to exercise his responsibility to be accurate, logical, fair, and reasonable in the use of facts and evidence. They may further note that if his view is wrong but acted on nonetheless, the consequences will be serious, and Mr. Limbaugh provides no warranty for buyers of his opinions that subsequently turn out to be defective.

Mr. Limbaugh is not alone. Right-wing think tanks and an odd alliance of rapacious corporations, aggrieved landowners, and the philosophically befuddled are also undermining good science and sound public policy in the cause of exploitation. With no detectable humor or irony, one such group calls itself the "wise use movement." They are now hard at work infiltrating public school boards, state legislatures, and boardrooms throughout the United States. To meet them effectively and forthrightly, a constituency for the long run must be able to tell the difference between politicized science and honest disagreement and between errant nonsense and reasoned discourse.

A second and even more pervasive barrier to the creation of a constituency for the long run is the widespread tendency to deny the seriousness of our situation. There is no honest way around the reality that the big numbers having to do with population growth, disruption of the earth's biogeochemical cycles, species extinction, and the health of soils, forests, and water are running against us. No one of these is necessarily fatal to our prospects. Taken together, however, they point inescapably to the conclusion that we do not have much time to set things right if we are to avoid major traumas in the decades ahead. The momentum of big numbers is sweeping us toward a precipice, but the words, concepts, theories, and stories essential to comprehend our situation are not yet part of our political language or public mind-set.

A third force working against us is the widespread belief that citizenship requires little or nothing of us. The idea of cheap citizenship is founded on the theology of the lottery: that one does not reap what one sows. It follows, then, that one need not sow at all, and that reaping is only a matter of luck, chicanery, or happenstance, not hard work, skill, and obligation. The mind-set of cheap citizenship is attributable, in part, to decades of televised bamboozlement. Some of it reflects the lingering effects of self-indulgences past, notably those of the 1980s. But the idea that one can get something for nothing is also built into the modern mind, which believes in nothing quite so zealously as it does in the heroic power of technology to absolve us of ecological malfeasance and ineptitude.

Our inability to deal seriously with a rising deficit provides a textbook study of cheap citizenship. Almost everyone agrees that the situation is serious, but few are willing to give up anything to solve the problem. The philosophy of cheap citizenship likewise prevents any serious discussion about paying the full costs of what we consume, including the costs of biotic impoverishment.

Cheap citizenship is, of course, an oxymoron. Real citizenship, political or ecological, is hard won and never more than temporarily won at that. Real citizens pay their bills, exercise foresight, assign costs and benefits fairly, work hard at maintaining their communities, and are willing to sacrifice when necessary and consider doing so a privilege. All of this is to say that authentic citizenship—political and ecological—is not cheap, but it is, sooner or later, less costly by far than dereliction and counterfeit citizenship.

Fourth, the manner in which we receive information about the world often works against the creation of a constituency for the long run. For example, in its quest for high ratings, television news has become little more than entertainment. If the end of the world arrives anytime soon, news of it will be brought to us by all of the major networks, each competing for the highest Nielsen ratings. The sponsors will include many of the corporations whose various activities made the event possible. We, the viewing public, will be dazzled by the graphics, amazed by the artistry of the advertisements, charmed by blow-dry reporters, sobered by the gravity of various commentators, and overcome by the spectacle of it all— entertained right into oblivion.

To entertain, it is necessary to create conflict and dramatic tension, often where none exists. Why is it, for example, that E. O. Wilson's views on species extinction are "balanced" by counterarguments of economist Julian Simon, whose scientific credentials are nil? News as entertainment necessarily avoids arduous details and thoughtful analysis of complex issues that might tax minds whittled down to fit 30-second sound bites. The exact rate at which species are being lost, for example, cannot be known with much precision. There is, accordingly, room for honest disagreement about causes, rates, consequences, and appropriate remedies. But to achieve "balance" by giving equal credence, say, to E. O. Wilson's views on the matter and those of Julian Simon is a bit like equating the views of the National Aeronautics and Space Administration with those of the Flat Earth Society about the shape of the planet (Stevens, 1991). The "balance" thus achieved tells us something about the range of opinions and perhaps a bit about the outer boundaries of the mind, but it neither clarifies nor informs. To the ecologically illiterate, such tripe only confuses already complex issues.

Similarly, recent reports of divergent opinions about trends in the average temperature of the earth were framed to appear as if scientists were in serious disagreement (Rensberger, 1993). Satellite recordings of temperatures four miles above the surface of the earth reveal no upward trend, while ground temperatures (until the eruption of Mt. Pinatubo) showed a sharp increase from 1980 through 1991. These are different and not necessarily conflicting data. To the average reader, however, the story appears to give yet one more reason to believe that scientists do not agree about the reality of global warming, hence one more reason to procrastinate (Rensberger, 1993). In the meantime real disagreements, including those about the larger risks and the ethics of our taking such risks for no good reason, go largely unmentioned.

❖ Upshot ❖

A constituency able and willing to fight for the long-term human prospect must be educated into existence. It must be scientifically literate enough to recognize politicized science for what it is. It must be courageous enough to face facts squarely. It must be committed enough to avoid the seductions of cheap citizenship. It must be intellectually alive enough to demand careful and thoughtful analysis of public problems. It must be able to tell the difference between ecological sense and nonsense. This will require, in Paul Kennedy's (1993) words, "nothing less than the re-education of humanity" (p. 331).

But there's the rub. What are schools, colleges, and universities doing to reeducate the citizenry or their own faculty, administrations, and trustees for that matter? The short answer is "not nearly enough," and in most cases the answer is "nothing at all." Even in this time of ecological concern, high schools, colleges, and universities continue to turn out a large percentage of graduates who have no clue how their personal prospects are intertwined with the vital signs of the earth. How could this be? Dartmouth professor Noel Perrin (1992) believes it to be a failure of leadership: "Neither the trustees nor the administration [of this or any other college or university] seems to believe that a crisis is coming" (p. B3). They comprehend the situation intellectually, Perrin believes, but they do not yet feel it at a gut level where action begins.

Among those familiar with education, few would disagree with such skepticism. To create a constituency for the long haul, we need farsighted leadership at all educational levels committed to making ecological literacy central to the debate about national educational goals and standards. That debate should be informed by the recognition that environmental education is not the same kind of education that enabled us to industrialize the earth. On the contrary, the kind of education we need begins with the recognition that the crisis of global ecology is first and foremost a crisis of values, ideas, perspectives, and knowledge, which makes it a crisis *of* education, not one *in* education.

There is an irreducible body of knowledge that all students should know, including how the earth works as a physical system, basic knowledge of ecology and thermodynamics, the vital signs of the earth, the essentials of human ecology, the natural history of their own region, and the kinds of knowledge that will enable them to restore natural systems and build ecologically resilient communities and economies. Accordingly, the reeducation of teachers, administrators, and boards of trustees must be a high priority. Those presuming to shape the minds that will shape the future must comprehend what the future requires of them.

SOURCES

Gore, A. 1992. *Earth in the Balance: Ecology and the Human Spirit.* Boston: Houghton Mifflin.

Kennedy, P. 1993. *Preparing for the Twenty-First Century.* New York: Random House.

Ornstein, R., and Ehrlich, P. 1989. *New World, New Mind.* New York: Doubleday.

Perrin, N. 1992, October 28. Colleges are Doing Pitifully Little to Protect the Environment. *The Chronicle of Higher Education, 39,* 10.

Rensberger, B. 1993, August 2–8. Blowing Hot and Cold on Global Warming. *Washington Post Weekly,* p. 38.

Stevens, W. 1991, August 20. Species Loss: Crisis or False Alarm? *The New York Times.*

Taubes, G. 1993, June 11. The Ozone Backlash. *Science, 260,* pp. 1580–1583.

PART FOUR

DESTINATIONS

A ND FOR what destination and for what destiny do we educate our children? For all of the fashionable talk about multiculturalism, the fact is that modern education has contributed greatly to the destruction of local cultures virtually everywhere. Locality has no standing in the modern curriculum. Abstractions, generalized knowledge, and technology do. Education has become a great homogenizing force undermining local knowledge, indigenous languages, and the self-confidence of placed people. It has become an adjunct to the commercial economy. It has hired itself out to the forces of growth and development, which as far as I can tell, is the effort to make the world safe for big capital. Taken as a whole, education has lacked the courage to ask itself what kind of world its graduates will inherit and what kind of world they will be prepared to build.

The essays in Part four suggest an alternative destiny. Whatever our powers as technological magicians, humankind, as Loren Eiseley (1970) expressed it, "lies under the spell of a greater and green enchantment" (p. 140). Something tugs at us that is not explained in theories of rational choice, self-interest, and economic maximization. Chapter 20 explores the hypothesis advanced by E. O. Wilson that this enchantment is an innate affinity for life that he calls "biophilia." The implications of that view have profound consequences for how we live in particular places, the subject of Chapter 21. Chapter 22 deals with an obstacle to living well in any place: dishonest accounting for the full costs of provisioning ourselves. The final chapter is a conjecture about the future balance between rural and urban areas in a world that will have to come to grips with limits of many kinds. I do not see limits ahead as necessarily a bad thing, however. To the contrary, the recognition of limits requires maturity and is the foundation for a better and more durable civilization.

SOURCES

Eiseley, L. 1970. *The Invisible Pyramid*. New York: Scribners.

CHAPTER TWENTY

Love It or Lose It:
The Coming Biophilia Revolution

I have set before you life and death, blessing and cursing:
therefore choose life, that both thou and thy seed may live.
— DEUTERONOMY 30:19

NATURE and I are two," filmmaker Woody Allen once said, and apparently the two have not gotten together yet (Lax, 1992, pp. 39–40). Allen is known to take extraordinary precautions to limit bodily and mental contact with rural flora and fauna. He does not go in natural lakes, for example, because "there are live things in there." The nature Allen does find comfortable is that of New York City, a modest enough standard for wildness.

Allen's aversion to nature, what can be called biophobia, is increasingly common among people raised with television, Walkman radios attached to their heads, and video games and living amidst shopping malls, freeways, and dense urban or suburban settings where nature is permitted tastefully, as decoration. More than ever we dwell in and among our own creations and are increasingly uncomfortable with nature lying beyond our direct control. Biophobia ranges from discomfort in "natural" places to active scorn for whatever is not manmade, managed, or air-conditioned. Biophobia, in short, is the culturally acquired urge to affiliate with technology, human artifacts, and solely with human interests regarding the natural world. I intend the word broadly to include as well those who regard nature "objectively" as nothing more than "resources" to be used any way the favored among the present generation see fit.

Is biophobia a problem as, say, misanthropy or sociopathy, or is it merely a personal preference; one plausible view of nature among many? Is it OK that Woody Allen feels little or no sympathy or kinship with nature? Does it matter that a growing number of other people do not like it or like it only in the abstract as nothing more than resources to be managed or as television nature specials? Does it matter that we are increasingly separated from the conditions of nature? If these things do matter, how do they matter and why? And why have so many come to think that the created world is inadequate? Inadequate to what and for what?

At the other end of the continuum of possible orientation toward nature is "biophilia," which E. O. Wilson (1984) has defined as "the urge to affiliate with other forms of life" (p. 85). Erich Fromm (1973) once defined it more broadly as "the passionate love of life and of all that is alive" (pp. 365–366). Both agree, however, that biophilia is innate and a sign of mental and physical health. To what extent are our biological prospects and our sanity now dependent on our capacity for biophilia? To that degree it is important that we understand how biophilia comes to be, how it prospers, what competencies and abilities it requires of us, and how these are to be learned.

Biophilia is not all that tugs at us. The affinity for life or biophilia competes with other drives and affinities, including biophobia disguised beneath the abstractions and presumptions of progress found in economics, management, and technology. Whatever is in our genes, then, the affinity for life is now a choice we must make. Compared with earlier cultures, our distinction lies in the fact that technology now allows us to move much further toward total domination of nature than ever before. Serious and well-funded people talk about reweaving the fabric of life on earth through genetic engineering and nanotechnologies, others talk of leaving the earth altogether for space colonies, and still others talk of reshaping human consciousness to fit "virtual reality." If we are to preserve a world in which biophilia can be expressed and can flourish, we will have to decide to make such a world.

❖ The Origins and Consequences of Biophobia ❖

In varying degrees humans have always modified their environments. I am persuaded that they generally have intended to do so with decorum and courtesy toward nature—not always and everywhere to be sure, but mostly. On balance, the evidence further suggests that biophilia or something close to it was woven throughout the myths, religions, and mindset of early humankind, which saw itself as participating with nature. In Owen Barfield's words, people once felt "integrated or mortised into" the world in ways that we do not and perhaps cannot (Barfield, 1957, p. 78). Technology, primitive by our standards, set limits on what tribal cultures could do to the world, while their myths, superstitions, and taboos constrained what they thought they ought to do. But I do not think that early humans *chose* biophilia, if for no other reason than that there was no choice to be made. And those tribes and cultures that were biophobic or incompetent toward nature passed into oblivion through starvation and disease (Diamond, 1992, pp. 317–338).

Looking back across that divide, I think it is evident that tribal cultures possessed an ecological innocence of sorts because they did not have the possibilities or the knowledge given to us. We, in contrast, must choose between biophobia and biophilia because science and technology have given us the power to destroy so completely as well as the knowledge to understand the consequences of doing so. The divide was not a sharp break but a kind of slow tectonic shift in perception and attitudes that widened throughout the late Middle Ages to the present. What we call "modernization" represented dramatic changes in how we regard the natural world and our role in it. These changes are now so thoroughly ingrained in us that we can scarcely conceive of any other manner of thinking. But crossing this divide first required us to discard the belief that the world was alive and worthy of respect, if not fear. To dead matter, we owe no obligations. Second, it was necessary to distance ourselves from animals who were transformed by Cartesian alchemy into mere machines. Again, no obligations or pity are owed to machines. In both cases, use is limited only by usefulness. Third, it was necessary to quiet whatever remaining sympathy we had for nature in favor of "hard" data that could be weighed, measured, counted, and counted on to make a profit. Fourth, we needed a reason to join power, cash, and knowledge in order to transform the world into more useful forms. Francis Bacon provided the logic, and the evolution of government-funded research did the rest. Fifth, we required a philosophy of improvement and found it in the ideology of perpetual economic growth, now the central mission of governments everywhere. Sixth, biophobia required the sophisticated cultivation of dissatisfaction, which could be converted into mass consumption. The advertising industry and the annual style change were invented.

For these revolutions to work, it was necessary that nature be rendered into abstractions and production statistics of board feet, tons, barrels, and yield. It was also necessary to undermine community, especially the small community, where attachment to place might grow and with it resistance to crossing the divide. Finally it was necessary to convert politics into the pursuit of material self-interest and hence render people impotent as citizens and unable to talk of larger and more important things.

To this point the story is well known, but it is hardly finished. Genetic engineers are busy remaking the fabric of life on earth. The development of nanotechnologies—machines at the molecular level—create possibilities for good and evil that defy prediction. How long will it be until the genetic engineers or nanotechnologists release an AIDS-like virus? One can only guess. But even those promoting such technologies admit that they "carry us toward unprecedented dangers . . . more potent than nuclear weapons" (Drexler, 1987, p. 174). And immediately ahead is the transformation of human consciousness brought on by the conjunction of neuroscience and computers in machines that will simulate whatever reality we choose. What happens to the quality of human experience or to our politics when cheap and thoroughgoing fantasy governs our mental life? In each case, untransformed nature pales by comparison. It is clumsy, inconvenient, flawed, and difficult to move or rearrange. It is slow. And it cannot be converted to mass dependence and profits so easily.

Beneath each of these endeavors lies a barely concealed contempt for unaltered life and nature, as well as contempt for the people who are expected to endure the mistakes, purchase the results, and live with the consequences, whatever those may be. It is a contempt disguised by terms of bamboozlement, like *bottom line, progress, needs, costs and benefits, economic growth, jobs, realism, research,* and *knowledge,* words that go undefined and unexamined. Few people, I suspect, believe "in their bones" that the net results from all of this will be positive, but most feel powerless to stop what seems to be so inevitable and unable to speak what is so hard to say in the language of self-interest.

The manifestation of biophobia, explicit in the urge to control nature, has led to a world in which it is becoming easier to be biophobic. Undefiled nature is being replaced by a defiled nature of landfills, junkyards, strip mines, clear-cuts, blighted cities, six-lane freeways, suburban sprawl, polluted rivers, and superfund sites, all of which deserve our phobias. Ozone depletion, meaning more eye cataracts and skin cancer, does give more reason to stay indoors. The spread of toxic substances and radioactivity does mean more disease. The disruption of natural cycles and the introduction of exotic species has destroyed much of the natural diversity that formerly graced our landscapes. Introduced blights and pests have or are destroying American chestnuts, elms, maples, dogwoods, hemlocks, and ashes. Global warming will degrade the flora and fauna of familiar places (Peters and Myers, 1991–1992, pp. 66–72). Biophobia sets into motion a vicious cycle that tends to cause people to act in such a way as to undermine the integrity, beauty, and harmony of nature, creating the very conditions that make the dislike of nature yet more probable.

Even so, is it OK that Woody Allen, or anyone else, does not like nature? Is biophobia merely one among a number of equally legitimate ways to relate to nature? I do not think so. First, for every "biophobe" others have to do that much more of the work of preserving, caring for, and loving the nature that supports biophobes and biophiliacs alike. Economists call this the "free-rider problem." It arises in every group, committee, or alliance when it is possible for some to receive all of the advantages of membership while doing none of the work necessary to create those advantages. Environmental free riders benefit from others' willingness to fight for the clean air that they breathe, the clean water that they drink, the preservation of biological diversity that sustains them, and the conservation of the soil that feeds them. But they lift not a finger. Biophobia is not OK because it does not distribute fairly the work of keeping the earth or any local place.

Biophobia is not OK for the same reason that misanthropy and sociopathy are not OK. We recognize these as the result of deformed childhoods that create unloving and often violent adults. Biophobia in all of its forms similarly shrinks the range of experiences and joys in life in the same way that the inability to achieve close and loving relationships limits a human life. E. O. Wilson (1984) put it this way:

> People can grow up with the outward appearance of normality in an environment largely stripped of plants and animals, in the same way that passable looking monkeys can be raised in laboratory cages and cattle fattened in feeding bins. Asked if they were happy, these people would probably say yes. Yet something vitally important would be missing, not merely the knowledge and pleasure that can be imagined and might have been, but a wide array of experiences that the human brain is peculiarly equipped to receive. (p. 118)

Can the same be said of whole societies that distance themselves from animals, trees, landscapes, mountains, and rivers? Is mass biophobia a kind of collective madness? In time I think we will come to know that it is.

Biophobia is not OK because it is the foundation for a politics of domination and exploitation. For our politics to work as they now do, a large number of people must not like any nature that cannot be repackaged and sold back to them. They must be ecologically illiterate and ecologically incompetent, and they must believe that this is not only inevitable but desirable. Furthermore, they must be ignorant of the basis of their dependency. They must come to see their bondage as freedom and their discontents as commercially solvable problems. The drift toward a biophobic society, as George Orwell and C. S. Lewis foresaw decades ago, requires the replacement of nature and human nature by technology and the replacement of real democracy by a technological tyranny now looming on the horizon.

These are reasons of self-interest: It is to our advantage to distribute the world's work fairly, to build a society in which lives can be lived fully, and to create an economy in which people participate knowledgeably. There is a further argument against biophobia that rests not on our self-interest, but on our duties. Finally, biophobia is not OK because it violates an ancient charge to replenish the earth. In return for our proper use, the earth is given to humankind as a trust. Proper use requires gratitude, humility, charity, and skill. Improper use begins with ingratitude and disparagement and proceeds to greed, abuse, and violence. We cannot forsake the duties of stewardship without breaking another trust with those who preceded us and with those who will follow.

Biophobia is certainly more complex than I have described it. One can be both biophobic and a dues-paying member of the Sierra Club. It is possible to be nature averse but still "like" the idea of nature as an abstraction. Moreover, it is possible to adopt the language and guise of biophilia and do a great deal of harm to the earth, knowingly or unknowingly. In other words, it is possible for us to be inconsistent, hypocritical, and ignorant of what we do.

But is it possible for us to be neutral or "objective" toward life and nature? I do not think so. On closer examination, what often passes for neutrality is nothing of the sort but rather the thinly disguised self-interest of those with much to gain financially or professionally. For those presuming to wear the robes of objectivity, the guise, in Abraham Maslow's

(1966) words, is often "a defense against being flooded by the emotions of humility, reverence, mystery, wonder, and awe" (p. 139). Life ought to excite our passion, not our indifference. Life in jeopardy ought to cause us to take a stand, not retreat into a spurious neutrality. Furthermore, it is a mistake to assume that commitment precludes the ability to think clearly and to use evidence accurately. To the contrary, commitment motivates intellectual clarity, integrity, and depth. We understand this in other realms quite well. When the chips are down, we do not go to physicians who admit to being neutral about the life and death of their patients. Nor when our hide is at stake do we go to lawyers who profess "objective" neutrality between justice and injustice. It is a mistake to think that matters of environment and life on earth are somehow different. They are not, and we cannot in such things remain aloof or indifferent without opening the world to demons.

❖ Biophilia ❖

We relate to the environment around us in different ways, with differing intensity, and these bonds have different sources. At the most common level, we learn to love what has become familiar. There are prisoners who prefer their jail cell to freedom; city dwellers, like Woody Allen, who shun rural landscapes or wilderness; and rural folk who will not set foot in the city. Simply put, we tend to bond with what we know well. Geographer Yi-Fu Tuan (1974) described this bonding as "topophilia," which includes "all of the human being's affective ties with the material environment" (p. 93). Topophilia is rooted less in our deep psychology than it is in our particular circumstances and experiences. It is closer to a sense of habitat that is formed out of the familiar circumstances of everyday living than it is a genuine rootedness in the biology and topography of a particular place. It is not innate, but acquired. New Yorkers have perhaps a greater sense of topophilia or habitat than do residents of Montana. But Montanans are more likely to feel kinship with sky, mountains, and trout streams. Both, however, tend to be comfortable with what has become habitual and familiar.

E. O. Wilson (1984) suggested a deeper source of attachment that goes beyond the particularities of habitat. "We are," he argues, "a biological species [that] will find little ultimate meaning apart from the remainder of life" (p. 112). We are bound to living things by what Wilson described as an innate urge to affiliate, or "biophilia," which begins in early childhood and "cascades" into cultural and social patterns. Biophilia is inscribed in the brain itself, expressing tens of thousands of years of evolutionary experience. It is evident in our preference for landscapes that replicate the savannas on which mind evolved: "Given a completely free choice, people gravitate statistically toward a savanna-like environment" (Wilson, 1984, p. 115). Removed to purely artificial environments and deprived of "beauty and mystery," the mind "will drift to simpler and cruder configurations," which undermine sanity itself (Wilson, 1984, p. 118). Still, biophilia competes with what Wilson describes as the "audaciously destructive tendencies of our species" that seem also to have "archaic biological origins" (p. 121). Allowing these tendencies free rein to destroy the world "in which the brain was assembled over millions of years" is, Wilson has argued, "a risky step."

A third possibility is that at some level of alertness and maturity, we respond with awe to the natural world independent of any instinctual conditioning. "If you study life deeply," Albert Schweitzer (1969) once wrote, "its profundity will seize you suddenly with dizziness" (p. 115). He described this response as "reverence for life" arising from the awareness of the unfathomable mystery of life itself. (The German word Schweitzer used, *Ehrfurcht*, implies greater awe than is implied by the English word *reverence*.) Reverence for life, I think, is akin to what Rachel Carson (1965/1987) meant by "the sense of wonder." But for Schweitzer (1972) reverence for life originated in large measure from the intellectual contemplation of the world: "Let a man once begin to think about the mystery of his life and the links which connect him with the life that fills the world, and he cannot but bring to bear upon his own life and all other life that comes within his reach the principle of Reverence for Life" (p. 231). Schweitzer regarded reverence for life as the only possible basis for a philosophy on which civilization might be restored from the decay he saw throughout the modern world. "We must," he wrote, "strive together to attain to a theory of the universe affirmative of the world and of life" (Schweitzer, 1972, p. 64).

We have reason to believe that this intellectual striving is aided by what is already innate in us and may be evident in other creatures. No less an authority than Charles Darwin believed that "all animals feel wonder" (Darwin, 1977, p. 450). Primatologist Harold Bauer once observed a chimpanzee lost in contemplation by a spectacular waterfall in the Gombe Forest Reserve in Tanzania. Contemplation finally gave way to "pant-hoot" calls while the chimp ran back and forth drumming on trees with its fists (Konner, 1982, p. 431). No one can say for certain what this behavior means, but it is not farfetched to see it as a chimpanzee version of awe and ecstasy. Jane Goodall and others have described similar behavior. It would be the worst kind of anthropocentrism to dismiss such accounts in the belief that the capacity for biophilia and awe is a human monopoly. In fact it may be that we have to work at it harder than other creatures. Joseph Wood Krutch (1991), for one, believed that for birds and other creatures "joy seems to be more important and more accessible than it is to us" (p. 227). And not a few philosophers have agreed with Abraham Heschel (1990) that "as civilization advances, the sense of wonder almost necessarily declines" (p. 37).

Do we, with all of our technology, retain a built-in affinity for nature? I think so, but I know of no proof that would satisfy skeptics. If we do have such an innate sense, we might nevertheless conclude from the damage that we have done to the world that biophilia does not operate everywhere and at all times. It may be, as Erich Fromm (1973) argued, that biophilia can be dammed up or corrupted and can subsequently appear in other, more destructive forms:

> Destructiveness is not parallel to, but the alternative to biophilia. Love of life or love of the dead is the fundamental alternative that confronts every human being. Necrophilia grows as the development of biophilia is stunted. Man is biologically endowed with the capacity for biophilia, but psychologically he has the potential for necrophilia as an alternative solution. (p. 366)

We also have reason to believe that people can lose the sense of biophilia. For example, in his autobiography, Darwin (1958) admitted that "fine scenery . . . does not cause me the exquisite delight which it formerly did" (p. 54). It is also possible that entire societies can lose the capacity for love of any kind. When the Ik tribe in northern Uganda was forcibly moved from its traditional hunting grounds into a tiny reserve, their world, as Colin Turnbull (1972) expressed it, "became something cruel and hostile," and they "lost whatever love they might once have had for their mountain world" (pp. 256, 259). The love for their place the Ik people may have once felt was transmuted into boredom and a "moody distrust" of the world around them and matched by social relations that Turnbull described as utterly loveless, cruel, and despicable. The Ik are a stark warning to us that the ties to life and to each other are more fragile than some suppose and, once broken, are not easily repaired or perhaps cannot be repaired at all.

Much of the history of the twentieth century offers further evidence of the fragility of biophilia and of philia. Ours is a time of unparalleled human violence and unparalleled violence toward nature. This is the century of Auschwitz and the mass extinction of species, nuclear weapons, and exploding economic growth.

Even if we could find no evidence of a lingering human affinity or affection for nature, however, humankind is now in the paradoxical position of having to learn altruism and selflessness, but for reasons of survival that are reasons of self-interest. In the words of Stephen Jay Gould (1991), "We cannot win this battle to save species and environments without forging an emotional bond between ourselves and nature as well—for we will not fight to save what we do not love" (p. 14). And if we do not save species and environments, we cannot save ourselves; we depend on those species and environments in more ways than we can possibly know. We have, in other words, "purely rational reasons" to cultivate biophilia (Wilson, 1984, p. 140).

Beyond our physical survival, there is still more at risk. The same Faustian urges that drive the ecological crisis also erode those qualities of heart and mind that constitute the essence of our humanity. Bertrand Russell (1959) put it this way:

> It is only in so far as we renounce the world as its lovers that we can conquer it as its technicians. But this division in the soul is fatal to what is best in man. . . . The power conferred by science as a technique is only obtainable by something analogous to the worship of Satan, that is to say, by the renunciation of love. . . . The scientific society in its pure form . . . is incompatible with the pursuit of truth, with love, with art, with spontaneous delight, with every ideal that men have hitherto cherished. (p. 264)

The ecological crisis, in short, is about what it means to be human. And if natural diversity is the wellspring of human intelligence, then the systematic destruction of nature inherent in contemporary technology and economics is a war against the very sources of mind. We have good reason to believe that human intelligence could not have evolved in a lunar landscape, devoid of biological diversity. We also have good reason to believe that the sense of awe toward the creation had a great deal to do with the origin of language and that early hominids *wanted* to talk, sing, and write poetry in the first place. Elemental things like flowing water, wind, trees, clouds, rain, mist, mountains, landscape, animals, changing seasons, the night sky, and the mysteries of the life cycle gave birth to thought and language. They continue to do so, but perhaps less exuberantly than they once did. For this reason I think it not possible to unravel natural diversity without undermining human intelligence as well. Can we save the world and anything like a human self from the violence we have unleashed without biophilia and reverence for the creation? All the arguments made by technological fundamentalists and by the zealots of instrumental rationality notwithstanding, I know of no good evidence that we can. We must choose, in Joseph Wood Krutch's (1991) words, whether "we want a civilization that will move toward some more intimate relation with the natural world or . . . one that will continue to detach and isolate itself from both a dependence upon and a sympathy with that community of which we were originally a part?" (p. 165). The writer of Deuteronomy had it right. Whatever our feelings, however ingenious our philosophies, whatever innate gravity tugs at us, we must finally choose between life and death, between intimacy and isolation.

❖ Biophilia: Eros to Agape ❖

We are now engaged in a great global debate about what it means to live "sustainably" on the earth. This word, however, is fraught with confusion, in large part because we are trying to define it before we have decided whether we want an intimate relation with nature or total mastery, as Krutch (1991) put it. We cannot know what sustainability means until we have decided what we intend to sustain and how we propose to do so. For some, sustainability means maintaining our present path of domination, only with greater efficiency. But were we to decide, in concurrence with Krutch and others, that we do want an intimate relation with nature, to take nature as our standard, what does that mean? We must choose along the continuum that runs between biophilia and biophobia and between intimacy and mastery, but how can we know when we have crossed over from one to the other? The choices are not always so simple, nor will they be presented to us so candidly. The options, even the most destructive, will be framed as life-serving, as necessary for a greater good someday, or as simply inevitable since "you can't stop progress." How, then, can we distinguish those things that serve life broadly and well from those that diminish it?

Biophilia is a kind of philia or love, but what kind? The Greeks distinguished three kinds of love: *eros*, meaning love of beauty or romantic love aiming to possess; *agape*, or sacrificial love, which asks nothing in return; and *philia*, or the love between friends. The first two of these reveal important parts of biophilia, which probably begins as eros but matures, if at all, as a form of agape. For the Greeks eros went beyond sensuous love to include creature needs for food, warmth, and shelter, as well as higher needs to understand, appreciate, and commune with nature (Bratton, 1992, p. 11). But eros aims no higher than self-fulfillment. Defined as an "innate urge," biophilia is eros, reflecting human desire and self-interest, including the interest in survival.

Biophilia as eros, however, traps us in a paradox. According to Susan Bratton (1992), "Without agape, human love for nature will always be dominated by unrestrained eros and distorted by extreme self-interest and material valuation" (p. 15). What we love only from self-interest, we will sooner or later destroy. Agape tempers our use of nature so that "God's providence is respectfully received and insatiable desire doesn't attempt to extract more from creation than it can sustain" (Bratton, 1992, p. 13). Agape enlarges eros, bringing humans and the creation together so that it is not possible to love either humanity or nature without also loving and serving the other. Agape in this sense is close to Schweitzer's description of "reverence for life," which calls us to transcend even the most enlightened calculations of self-interest. Wouldn't respect for nature do as well? I think not, and for the reason that it is just too bloodless, too cool, and too self-satisfied and aloof to cause us to do much to save species and environments. I am inclined to agree with Stephen Jay Gould that we will have to reach deeper.

What, then, do we know about deeper sources of motivation, including the ways in which eros is transformed into agape, and what does this reveal about biophilia? First, we know that the capacity for love of any kind begins early in the life and imagination of the child. The potential for biophilia possibly begins at birth, as Robert Coles once surmised, with the newborn infant being introduced to its place in nature (Coles, 1971). If so, the manner and circumstances of birth are more important than is usually thought. Biophilia is certainly evident in the small child's efforts to establish intimacy with the earth, like that of Jane Goodall, age two, sleeping with earthworms under her pillow (Montgomery, 1991, p. 28), or John Muir (1988), "reveling in the wonderful wildness" around his boyhood Wisconsin home (p. 43). If by some fairly young age, however, nature has not been experienced as a friendly place of adventure and excitement, biophilia will not take hold as it might have. An opportunity will have passed, and thereafter the mind will lack some critical dimension of perception and imagination.

Second, I think we know that biophilia requires easily and safely accessible places where it might take root and grow. For Aldo Leopold it began in the marshes and woods along the Mississippi River. For young E. O. ("Snake") Wilson (1984) it began in boyhood explorations of the "woods and swamps in a languorous mood . . . [forming] the habit of quietude and concentration" (pp. 86–92). The loss of places such as these is one of the uncounted costs of economic growth and urban sprawl. It is also a powerful argument for containing that sprawl and expanding urban parks and recreation areas.

Third, I think we can safely surmise that biophilia, like the capacity to love, needs the help and active participation of parents, grandparents, teachers, and other caring adults. Rachel Carson's (1987) relation with her young nephew caused her to conclude that the development of a child's sense of wonder required "the companionship of at least one adult who can share it, rediscovering with him the joy, excitement and mystery of the world we live in" (p. 45). For children the sense of biophilia needs instruction, example, and validation by a caring adult. And for adults, rekindling the sense of wonder may require a child's excitement and openness to natural wonders as well.

Fourth, we have every reason to believe that love and biophilia alike flourish mostly in good communitites. I do not mean necessarily affluent places. In fact, affluence often works against real community, as surely as does violence and utter poverty. By community I mean, rather, places in which the bonds between people and those between people and the natural world create a pattern of connectedness, responsibility, and mutual need. Real communities foster dignity, competence, participation, and opportunities for good work. And good communities provide places in which children's imagination and earthy sensibilities root and grow.

Fifth, we have it on good authority that love is patient, kind, enduring, hopeful, long-suffering, and truthful, not envious, boastful, insistent, arrogant, rude, self-centered, irritable, and resentful (I Corinthians 13).

For biophilia to work, I think it must have similar qualities. Theologian James Nash (1991) for example proposed six ecological dimensions of love: (1) beneficence, e.g., kindness to wild creatures; (2) other-esteem, which rejects the idea of possessing or managing the biosphere; (3) receptivity to nature, e.g., awe; (4) humility, by which is meant caution in the use of technology; (5) knowledge of ecology and how nature works; and (6) communion as "reconciliation, harmony, koinonia, shalom" between humankind and nature (pp. 139–161). I would add only that real love does not do desperate things, and it does not commit the irrevocable.

Sixth, I think we know with certainty that beyond some scale and level of complexity, the possibility for love of any sort declines. Beneficence, awe, reconciliation, and communion are not entirely probable attitudes for the poverty stricken living in overcrowded barrios. With 10 or 12 billion people on the earth, we will have no choice but to try to manage nature, even though it will be done badly. The desperate and the hungry will not be particularly cautious with risky technologies. Nor will the wealthy, fed and supplied by vast, complex global networks, understand the damage they cause in distant places they never see and the harm they do to people they will never know. Knowledge has its own limits of scale. Beyond some level of scale and complexity, the effects of technology, used in a world we cannot fully comprehend, are simply unknowable. When the genetic engineers and the nanotechnologists finally cause damage to the earth comparable to that done by the chemists who invented and so casually and carelessly deployed chlorofluorocarbons, they too will plead for forgiveness on the grounds that they did not know what they were doing.

Seventh, love, as Eric Fromm (1989) wrote, is an art, the practice of which requires "discipline, concentration and patience throughout every phase of life" (p. 100). The art of biophilia, similarly, requires us to use the world with disciplined, concentrated, and patient competence. To live and earn our livelihood means that we must "daily break the body and shed the blood of creation," in Wendell Berry's (1981) words. Our choice is whether we do so "knowingly, lovingly, skillfully, reverently . . . [or] ignorantly, greedily, clumsily, destructively" (p. 281). Practice of any art also requires forbearance, which means the ability to say no to things that diminish the object of love or our capacity to work artfully. And for the same reasons that it limits the exploitation of persons, forbearance sets limits on our use of nature.

Finally, we know that for love to grow from eros to agape, something like *metanoia*, or the "transformation of one's whole being" is necessary. Metanoia is more than a "paradigm change." It is a change, first, in our loyalties, affections, and basic character, which subsequently changes our intellectual priorities and paradigms. For whole societies, the emergence of biophilia as agape will require something like a metanoia that deepens our loyalty and affections to life and over time alters the character of our entire civilization.

THE BIOPHILIA REVOLUTION

"Is it possible," E. O. Wilson (1984) asked, "that humanity will love life enough to save it?" (p. 145). And if we do love life enough to save it, what is required of us? On one level the answer is obvious. We need to transform how and how rapidly we use the earth's endowment of land, minerals, water, air, wildlife, and fuels: an efficiency revolution that buys us some time. Beyond efficiency, we need another revolution that transforms our ideas of what it means to live decently and how little is actually necessary for a decent life: a sufficiency revolution. The first revolution is mostly about technology and economics. The second revolution is about morality and human purposes. The biophilia revolution is about the combination of reverence for life and purely rational calculation by which we will *want* to both be efficient and live sufficiently. It is about finding our rightful place on earth and in the community of life, and it is about citizenship, duties, obligations, and celebration.

There are two formidable barriers standing in our way. The first is the problem of denial. We have not yet faced up to the magnitude of the trap we have created for ourselves. We are still thinking of the crisis as a set of problems that are, by definition, solvable with technology and money. In fact we face a series of dilemmas that can be avoided only through wisdom and a higher and more comprehensive level of rationality than we have yet shown. Better technology would certainly help; however, our crisis is not fundamentally one of technology but one of mind, will, and spirit. Denial must be met by something like a worldwide ecological "perestroika," predicated on the admission of failure: the failure of our economics, which became disconnected from life; the failure of our politics, which lost sight of the moral roots of our commonwealth; the failure of our science, which lost sight of the essential wholeness of things; and the failures of all of us as moral beings, who allowed these things to happen because we did not love deeply and intelligently enough. The biophilia revolution must come as an ecological enlightenment that sweeps out the modern supersitition that we are knowledgeable enough and good enough to manage the earth and to direct evolution.

The second barrier standing in the way of the biophilia revolution is one of imagination. It is easier, perhaps, to overcome denial than it is to envision a biophilia-centered world and believe ourselves capable of creating it. We could get an immediate and overwhelming worldwide consensus today on the proposition "Is the world in serious trouble?" But we are not within a light-year of agreement on what to do about it. Confronted by the future, the mind has a tendency to wallow. For this reason we can diagnose our plight with laser precision while proposing to shape the future with a sledgehammer. Fictional utopias, almost without exception, are utterly dull and unconvincing. And the efforts to create utopias of either right or left have been monumental failures, leaving people profoundly discouraged about their ability to shape the world in accord with their highest values. And now some talk about creating a world that is sustainable, just, and peaceful! What is to be done?

Part of our difficulty in confronting the future is that we think of utopia on too grand a scale. We are not very good at comprehending things on the scale of whole societies, much less that of the planet. Nor have we been very good at solving the problems utopias are supposed to solve without imposing simplistic formulas that ride roughshod over natural and cultural diversity. Except for some anarchists, utopianism is almost synonymous with homogenization. Another part of the problem is the modern mind's desire for drama, excitement, and sexual sizzle, which explains why we do not have many bestselling novels about Amish society, arguably the closest thing to a sustainable society we know. How do we fulfill the need for meaning and variety while discarding some of our most cherished fantasies of domination? How do we cause the "change in our intellectual emphasis, loyalties, affections, and convictions," without which all else is moot? (Leopold, 1966, p. 246) When we think of revolution, our first impulse is to think of some grand political, economic, or technological change; some way to fix quickly what ails us. What ails us, however, is closer to home, and I suggest that we begin there.

THE RECOVERY OF CHILDHOOD : I began by describing biophilia as a choice. In fact it is a series of choices, the first of which has to do with the conduct of childhood and how the child's imagination is woven into a home place. Practically, the cultivation of biophilia calls for the establishment of more natural places, places of mystery and adventure where children can roam, explore, and imagine. This means more urban parks, more greenways, more farms, more river trails, and wiser land use everywhere. It means redesigning schools and campuses to replicate natural systems and functions. It means greater contact with nature during the school day but also unsupervised hours to play in places where nature has been protected or allowed to recover.

For biophilia to take root, we must take our children seriously enough to preserve their natural childhood. However, childhood is being impoverished and abbreviated, and the reasons sound like a curriculum in social pathology: too many broken homes and unloving marriages, too much domestic violence, too much alcohol, too many drugs, too many guns, too many things, too much television, too much idle time and permissiveness, too many off-duty parents, and too little contact with grandparents. Children are rushed into adulthood too soon, only to become childish adults unprepared for parenthood, and the cycle repeats itself. We will not enter this new kingdom of sustainability until we allow our children the kind of childhood in which biophilia can put down roots.

RECOVERING A SENSE OF PLACE: I do not know whether it is possible to love the planet or not, but I do know that it is possible to love the places we can see, touch, smell, and experience. And I believe, along with Simone Weil (1971), that rootedness in a place is "the most important and least recognized need of the human soul" (p. 43). The attempt to encourage biophilia will not amount to much if we fail to decide to reshape these kinds of places so that we might become deeply rooted. The second decision we must make, then, has to do with the will to rediscover and reinhabit our places and regions, finding in them sources of food, livelihood, energy, healing, recreation, and celebration. Whether one calls it "bioregionalism" or "becoming native to our places" it means deciding to relearn the arts that Jaquetta Hawkes (1951) once described as "a patient and increasingly skillful love-making that [persuades] the land to flourish" (p. 202). It means rebuilding family farms, rural villages, towns, communities, and urban neighborhoods. It means restoring local culture and our ties to local places, where biophilia first takes root. It means reweaving the local ecology into the fabric of the economy and life patterns while diminishing use of the automobile and our ties to the commercial culture. It means deciding to slow down, hence more bike trails, more gardens, and more solar collectors. It means rediscovering and restoring the natural history of our places. And, as Gary Snyder (1974) wrote, it means finding our place and digging in (p. 101).

EDUCATION AND BIOPHILIA: The capacity for biophilia can still be snuffed out by education that aims no higher than to enhance the potential for upward mobility, which has come to mean putting as much distance as possible between the apogee of one's career trajectory and one's roots. We should worry a good bit less about whether our progeny will be able to compete as a "world-class workforce" and a great deal more about whether they will know how to live sustainably on the earth. My third proposal, then, requires the will to reshape education in a way that fosters innate biophilia and the analytical abilities and practical skills necessary for a world that takes life seriously.

Lewis Mumford (1946) once proposed the local community and region as the "backbone of a drastically revised method of study" (pp. 150–154). The study of the region would ground education in the particularities of a specific place and would also integrate various disciplines around the "regional survey," which includes surveys of local soils, climate, vegetation, history, economy, and society. Mumford (1970b) envisioned this as an "organic approach to knowledge" that began with the "common whole—a region, its activities, its people, its configuration, its total life" (p. 385). The aim was "to educate citizens, to give them the tools of action" and to educate a people "who will know in detail where they live and how they live . . . united by a common feeling for their landscape, their literature and language, their local ways" (Mumford, 1970b, p. 386).

Something like the regional survey is required for the biophilia revolution. Education that supports and nourishes a reverence for life would occur more often out-of-doors and in relation to the local community. It would provide a basic competence in the kinds of knowledge that Mumford described a half century ago. It would help people become not only literate but ecologically literate, understanding the biological requisites of human life on earth. It would provide basic competence in what I have called the "ecological design arts," that is, the set of perceptual and analytic abilities, ecological wisdom and practical wherewithal essential to making things that fit in a world governed by the laws of ecology and thermodynamics.

A NEW COVENANT WITH ANIMALS: The biophilia revolution would be incomplete without our creating a new relationship with animals, one, in Barry Lopez's (1989) words, that "rise(s) above prejudice to a position of respectful regard toward everything that is different from ourselves and not innately evil" (p. 383). We need animals, not locked up in zoos, but living free on their own terms. We need them for what they can tell us about ourselves and about the world. We need them for our imagination and for our sanity. We need animals for what they can teach us about courtesy and what Gary Snyder (1990) called "the etiquette of the wild" (pp. 3–24). The human capacity for biophilia as agape will remain "ego-centric and partial" until it can also embrace creatures who cannot reciprocate (Mumford 1970a, p. 286). And needing animals, we will need to restore wild landscapes that invite them again.

A new covenant with animals requires that we decide to limit the human domain in order to establish their rights in law, custom, and daily habit. The first step is to discard the idea obtained from Rene Descartes that animals are only machines, incapable of feeling pain and to be used any way we see fit. Protecting animals in the wild while permitting confinement feeding operations and most laboratory use of animals makes no moral sense and diminishes our capacity for biophilia. In this, I think Paul Shepard (1993) is right: To recognize animals and wildness is to decide to admit deeper layers of our consciousness into the sunlight of full consciousness again.

THE ECONOMICS OF BIOPHILIA: The biophilia revolution will also require national and global decisions that permit life-centeredness to flourish at a local scale. Biophilia can be suffocated, for example, by the demands of an economy oriented toward accumulation, speed, sensation, and death. But economists have not written much about how an economy encourages or discourages love generally or biophilia in particular. As a result, not much thought has been given to the relationship between love and the way we earn our keep.

The transition to an economy that fosters biophilia requires a decision to limit the human enterprise relative to the biosphere. Some economists talk confidently of a five- or tenfold increase in economic activity over the next half century. But Peter Vitousek and his colleagues have shown that humans now use or coopt 40% of the net primary productivity from terrestrial ecosystems (Vitousek et al., 1986). What limits does biophilia set on the extent of the human enterprise? What margin of error does love require?

Similarly, in the emerging global economy, in which capital, technology, and information move easily around the world, how do we protect the people and the communities left behind? Now more than ever the rights of capital are protected by all the power money can buy. The rights of communities are protected less than ever. Consequently, we face complex decisions about how to protect communities and their stability on which biophilia depends.

BIOPHILIA AND PATRIOTISM: The decisions necessary to move us toward a culture capable of biophilia are, in the end, political decisions. But our politics, no less than our economy, has other priorities. In the name of "national security" or one ephemeral national "interest" or another we lay waste to our lands and to the prospects of our children. Politics of the worst sort has corrupted our highest values, becoming instead one long evasion of duties and obligations in the search for private or sectarian advantage. "Crackpot realists" tell us that this is how it has always been and must therefore always be: a view that marries bad history to bad morals.

Patriotism, the name we give to the love of one's country, must be redefined to include those things that contribute to the real health, beauty, and ecological stability of our home places and to exclude those that do not. Patriotism as biophilia requires that we decide to rejoin the idea of love of one's country to how and how well one uses the country. To destroy forests, soils, natural beauty, and wildlife in order to swell the gross national product, or to provide short-term and often spurious jobs, is not patriotism but greed.

Real patriotism requires that we weave the competent, patient, and disciplined love of our land into our political life and our political institutions. The laws of ecology and those of thermodynamics, which mostly have to do with limits, must become the foundation for a new politics. No one has expressed this more clearly than Vaclav Havel (1989): "We must draw our standards from our natural world. . . . We must honour with the humility of the wise the bounds of that natural world and the mystery which lies beyond them, admitting that there is something in the order of being which evidently exceeds all our competence" (p. 153). Elsewhere, Havel (1992) stated the following:

> Genuine Politics . . . is simply a matter of serving those around us: serving the community, and serving those who will come after us. Its deepest roots are moral because it is a responsibility, expressed through action, to and for the whole, a responsibility . . . only because it has a metaphysical grounding: that is, it grows out of a

conscious or subconscious certainty that our death ends nothing, because everything is forever being recorded and evaluated somewhere else, somewhere 'above us', in what I have called 'the memory of being'. . . . (p. 6)

❖ Conclusion ❖

Erich Fromm (1955) once asked whether whole societies might be judged sane or insane. After the World Wars, state-sponsored genocide, gulags, McCarthyism, and the "mutual assured destruction" of the twentieth century there can be no doubt that the answer is affirmative. Nor do I doubt that our descendants will regard our obsession with perpetual economic growth and frivolous consumption as evidence of theologically induced derangement. Our modern ideas about sanity, in large measure, can be attributed to Sigmund Freud, an urban man. And from the urban male point of view, the relationship between nature and sanity may be difficult to see and even more difficult to feel. Freud's reconnaissance of the mind stopped too soon. Had he gone further, and had he been prepared to see it, he might have discovered what Theodore Roszak (1992) called "the ecological unconscious," the repression of which "is the deepest root of collusive madness in industrial society" (p. 320). He may also have stumbled upon biophilia, and had he done so, our understanding of individual and collective sanity would have been on more solid ground.

The human mind is a product of the Pleistocene Age, shaped by wildness that has all but disappeared. If we complete the destruction of nature, we will have succeeded in cutting ourselves off from the source of sanity itself. Hermetically sealed amidst our creations and bereft of those of The Creation, the world then will reflect only the demented image of the mind imprisoned within itself. Can the mind doting upon itself and its creations be sane? Thoreau never would have thought so, nor should we.

A sane civilization that loved more fully and intelligently would have more parks and fewer shopping malls; more small farms and fewer agribusinesses; more prosperous small towns and smaller cities; more solar collectors and fewer strip mines; more bicycle trails and fewer freeways; more trains and fewer cars; more celebration and less hurry; more property owners and fewer millionaires and billionaires; more readers and fewer television watchers; more shopkeepers and fewer multinational corporations; more teachers and fewer lawyers; more wilderness and fewer landfills; more wild animals and fewer pets. Utopia? No! In our present circumstances this is the only realistic course imaginable. We have tried utopia and can no longer afford it.

SOURCES

Barfield, O. 1957. *Saving the Appearances*. New York: Harcourt Brace Jovanovich.

Berry, W. 1981. *The Gift of Good Land*. San Francisco: North Point Press.

Bratton, S. 1992, Spring. Loving Nature: Eros or Agape? *Environmental Ethics* 14, 1.

Carson, R. 1987. *The Sense of Wonder*. New York: Harper. (Original work published 1965.)

Coles, R. 1971. A Domain of Sorts. In S. Kaplan and R. Kaplan, eds., *Humanscape*. North Scituate, Mass.: Duxbury.

Darwin, C. 1958. *The Autobiography of Charles Darwin*. New York: Dover. (Original work published 1892.)

Darwin, C. 1977. *The Descent of Man*. New York: Modern Library. (Original work published 1871.)

Diamond, J. 1992. *The Third Chimpanzee*. New York: Harper.

Drexler, E. 1987. *Engines of Creation*. New York: Anchor Books.

Fromm, E. 1955. *The Sane Society*. New York: Fawcett Books.

Fromm, E. 1973. *The Anatomy of Human Destructiveness*. New York: Holt, Rinehart & Winston.

Fromm, E. 1989. *The Art of Loving*. New York: Harper.

Gould, S. 1991, September. Enchanted Evening. *Natural History*, p. 14.

Havel, V. 1989. *Living in Truth*. London: Faber & Faber.

Havel, V. 1992. *Summer Meditations*. New York: Knopf.

Hawkes, J. 1951. *A Land*. New York: Random House.

Heschel, A. 1990. *Man is not Alone*. New York: Farrar, Straus & Giroux.

Konner, M. 1982. *The Tangled Wing*. New York: Holt, Rinehart & Winston.

Krutch, J. 1991. *The Great Chain of Life*. Boston: Houghton Mifflin.

Lax, E. 1992. *Woody Allen: A Biography*. New York: Vintage.

Leopold, A. 1966. *A Sand County Almanac*. New York: Ballantine. (Original work published 1949.)

Lopez, B. 1989. Renegotiating the Contracts. In T. Lyon, ed., *This Incomperable Lande*. Boston: Houghton Mifflin.

Maslow, A. 1966. *The Psychology of Science*. Chicago: Gateway.

Montgomery, S. 1991. *Walking with the Great Apes*. Boston: Houghton Mifflin.

Muir, J. 1988. *The Story of My Boyhood and Youth*. San Francisco: Sierra Club.

Mumford, L. 1946. *Values for Survival*. New York: Harcourt and Brace.

Mumford, L. 1970a. *The Conduct of Life*. New York: Harcourt Brace Jovanovich.

Mumford, L. 1970b. *The Culture of Cities*. New York: Harcourt Brace Jovanovich.

Nash, J. 1991. *Loving Nature*. Nashville: Abingdon.

Peters, R., and Myers, J. P. 1991–1992. Preserving Biodiversity in a Changing Climate. *Issues in Science and Technology*, 8, 2.

Roszak, T. 1992. *The Voice of the Earth*. New York: Simon & Schuster.

Russell, B. 1959. *The Scientific Outlook*. New York: Norton.

Schweitzer, A. 1969. *Reverence for Life*. New York: Pilgrim Press.

Schweitzer, A. 1972. *Out of My Life and Thought*. New York: Holt, Rinehart & Winston.

Shepard, P. 1993. On Animal Friends. In S. Kellert and E. O. Wilson, eds., *The Biophilia Hypothesis*. Washington, DC: Island Press.

Shepard, P., and Sanders, B. 1992. *The Sacred Paw*. New York: Viking.

Snyder, G. 1974. *Turtle Island*. New York: New Directions.

Snyder, G. 1990. *The Practice of the Wild*. San Francisco: North Point Press.

Tuan, Y. 1974. *Topophilia*. New York: Columbia University Press.

Turnbull, C. 1972. *The Mountain People*. New York: Simon & Schuster.

Vitousek, P., et al. 1986, June. Human Appropriation of the Products of Photosynthesis. *Bioscience*, 36, 6.

Weil, S. 1971. *The Need for Roots*. New York: Harper.

Wilson, E. O. 1984. *Biophilia*. Cambridge: Harvard University Press.

CHAPTER TWENTY-ONE

A World That Takes Its Environment Seriously

Find your place on the planet, dig in, and take responsibility from there.

— GARY SNYDER

❖ Recollections ❖

I grew up in a small town amidst the rolling hills and farms of western Pennsylvania. As towns go it wasn't much different from hundreds of others throughout the United States. There was a main street with shops and stores, a funeral parlor or two, four churches, a small liberal arts college, and perhaps two thousand residents give or take. It was a "dry" town filled with serious and hard-working Protestants and a disconcertingly large number of retired preachers and missionaries. It was not a place that quickly welcomed Elvis and rock and roll. The prevailing political sensibilities were sober and overwhelmingly Republican of the Eisenhower sort. The town would have seemed stuffy and parochial to a Sherwood Anderson or a Theodore Drieser. And it probably was. By the standards of the 1990s, the town, the college, and its residents would have failed even the most lax certification for political correctness. It was a man's world, neither multicultural nor multiracial. The sexual revolution lay ahead. And almost everyone who was anyone in town bought without question the assumptions of mid-century America about our inherent virtue, economic progress, communism, and technology. J. Edgar Hoover was a hero. Boys were measured for manhood on the baseball diamond or the basketball court. It was also a place, like most others, in transition from one kind of economy to another.

Typical of most small towns, the main street of New Wilmington, Pennsylvania, still reflected bits of the nineteenth-century agrarian economy. There was, for example, a dilapidated and unused livery stable behind the main street where a funeral parlor parked a hearse. On Main Street, Mr. Meeks operated his watch repair shop and Mr. Fusco had his shoe repair shop. There were locally owned and operated businesses, including two grocery stores, a hardware and plumbing store, a good bakery, an electronics/appliance store, a dairy store, a bank, a dry goods store, a magazine and tobacco shop, a movie theater, a building supply store, and a butcher shop. The train station was located two blocks from main street. A half mile to the south a local entrepreneur operated a tool-making plant. A quarter mile beyond lay the town dump on the banks of Neshannock Creek.

The small-town, repair-and-reuse economy was predominantly locally owned and operated. My mother bought groceries from the store on Main Street. She bought vegetables from local farmers, including the Amish who went door to door selling everything from farm fresh eggs to maple syrup. Milk was delivered daily in returnable glass bottles by a locally owned dairy company. Soda pop also came in returnable glass bottles from a bottling plant eight miles distant. Broken machinery could be repaired in town. Dull saws could still be sharpened for a dime. Hand-me-down clothing was standard, and as the youngest I was the last stop for lots of items. And some of the best Christmas presents I ever received were made by hand.

The forces that would undermine that sheltered world of small-town, mid-century America were on the march. But I knew nothing of these as I joined the great exodus of self-assured and expectant young people leaving their hometowns for some other place thought to offer greater opportunity and more excitement. Few of us could say with certainty why we were going or where we were headed other than that it was somewhere else. Nor could we have said what we were leaving behind.

Looking back, I can see that even then things were changing as the larger industrial economy began to undermine local economies nearly everywhere. We bought our first television set the same year that Congress passed the Interstate Highway Act. I recall the lights on the big shovel at the strip mine across the valley burning into the night. The contractor for whom I worked in the summer went out of business shortly after I graduated from college. The farmer who gave me part-time employment, and was thought to be the most progressive in the county, went bankrupt in 1975. He was not alone. People in New Wilmington now buy their milk in plastic jugs from interstate dairy cooperatives. The local bottling plant disappeared and with it the practice of returning bottles to the store. The nearby industrial cities of New Castle and Youngstown, Ohio, which I knew as busy and thriving places, are now mostly derelict and abandoned, as are other cities in what was once a blue-collar, industrial corridor stretching from Pittsburgh to Cleveland. Interstate highways to the north and east of town now slash across what was once farm country. Tourism is the main economic hope. Crime, I hear, is a growing problem.

❖ Large Numbers ❖

In the 30 years since the class of 1961 set out to find its way, world population grew from 3.2 billion to 5.5 billion; approximately 120 billion tons of carbon dioxide were emitted to the atmosphere mostly from the combustion of fossil fuels; perhaps a tenth of the life forms on the earth disappeared in that time; a quarter of the world's rain forests were cut down; half or more of the forests in Europe were damaged by acid rain; careless farming and development caused the erosion of some 600 billion tons of topsoil worldwide; and the ozone shield was severely damaged. Before the class of 1961 is just a faint memory, the earth may be 2°C to 3°C hotter, with consequences we can barely imagine; world population will be 8–9 billion; perhaps 25% of the earth's species will have disappeared; and humans will have turned an area roughly equivalent to the size of the United States into desert. Something of earth-shattering importance went wrong in our lifetime, and we were prepared neither to see it nor to avoid complicity in it.

❖ Hindsight ❖

Looking back with more or less 20/20 hindsight, I believe that amidst all of the many good things in my town, there were three things missing, which bear on the issues implied in the title of this chapter. First, and most obvious, we were taught virtually nothing of ecology, systems, and interrelatedness. But neither were many others. This was a blind spot for a country determined to grow and armed with the philosophy of economic improvement. As a consequence we knew little of our ecological dependencies or, for that matter, our own vulnerabilities. The orchard beside our house was drenched with pesticides every spring and summer, and we never objected. The blight of nearby strip mines grew year by year, and we saw little wrong with that either.

We grew up in a bountiful region, which was virtually opaque to us. In school I learned about lots of other places, but I did not learn much about my own. We were not taught to think about how we lived in relation to where we lived. The Amish farms nearby, arguably the best example we have of a culture that fits its locale, were regarded as a quaint relic of a bygone world that had nothing to offer us. There was no course in high school or the local college on the natural history of the area. To this day, little has been written about the area as a bioregion. So we grew up mostly ignorant of the biological and ecological conditions in which we lived and what these required of us.

I finished high school the year before publication of Rachel Carson's (1962) *Silent Spring* but not before the projections of U.S. oil production by M. King Hubbert (1957), and some of the best writings of Lewis Mumford, Paul Sears, Fairfield Osborn, William Vogt, and earlier writings of John Muir, John Burroughs, George Perkins Marsh, and Henry David Thoreau. Our teachers and mentors had been through both the dust bowl and the Depression, but it was the latter that affected them most and that fact could not help but affect us. Almost by osmosis we absorbed the purported lessons of economic hardship, but not those of ecological collapse, which can also lead to privation and economic failure. When it came time to rebel, we did so over such things as "lifestyle" and music. But we in the classs of 1961 had no concept of enough or any reason to think that limits of any sort were important. Inadequate though it was, we did have an economic philosophy, but we had no articulate or ecologically solvent view of nature. We were sent out into the world armed with a creed of progress but had scarcely a clue about our starting point or how to "find our place and dig in." And none of us in 1961 would have had any idea of what those words meant.

Looking back, I can see a second missing element. On one hand I recall no skepticism or even serious discussion about technology. On the other, the college-bound students were steered into academic courses and away from vocational courses. As a result the upwardly mobile became both technologically illiterate and technologically incompetent. All the while there was a "what will they think of next" kind of naivete reinforced by advertisers' hawking messages about "living better electrically" and "progress as our most important product," which we accepted without much thought. We were good at detecting the benefits of technology in parts per billion and did not see until much later what it would cost us. Nor could we see the web of dependencies that was beginning to entrap us. The same "they" who would somehow figure it all out were taking the things that Americans once did for themselves as competent people, citizens, and neighbors and selling them back at a good markup. We were turned out into the world with the intellectual equivalent of a malfunctioning immune system, unable to think critically about technology. If we read Faust at all, we read it as a fable, not as a prophecy.

Third, had we known our place better, and had we been ecologically literate and technologically savvy, we still would have lacked the political wherewithal to be better stewards of our land and heritage. Our version of small-town, flag-waving patriotism was disconnected from the tangible things of livelihood and location, soils and stewardship. We mistook the large abstractions of nationalism, flag, and Presidential authority for patriotism. Accordingly, we were vulnerable to the chicanery of Joe McCarthy and J. Edgar Hoover, and to Lyndon Johnson's lies about Vietnam, Richard Nixon's lies about nearly everything, and Ronald Reagan's fantasies about "morning in America."

My classmates and I are, I think, typical of most Americans born and raised in the middle decades of the twentieth century. Ours has been a time of cheap energy, economic and technological optimism, lots of patriotic huffing and puffing, and "auto-mobility." We are movers and we move on average eight to ten times in a lifetime. We were educated to be competent in an industrial world and incompetent in any other. We did not much question the values and assumptions of the industrial "paradigm" or those underlying notions of progress. Those beliefs were givens. We were turned out into the world, vulnerable to whatever economic, technological, or even political changes would be thrust upon us, as long as they were said to be economically necessary or simply inevitable. We were not taught to question the physical, biological, and psychological reordering of the world taking place all around us. Nor were we enabled to see it for what it was.

New Wilmington, Pennsylvania, is still a nice town. Having little industry, it has not suffered the rusting fate of the nearby industrial cities. It has also been spared some of the uncontrolled growth that has desecrated many other regions. Housing developments outside town, though, are now filling up what was once good farmland. Aside from the Amish, the local farm economy is a shadow of what it once was. The effects of acid rain are beginning to show on trees. To make ends meet, the region is increasingly dependent on tourism. New Wilmington, like most small towns, is an island at the mercy of decisions made elsewhere. It has been spared mostly because no one noticed it or thought it a place likely to be profitable enough for an interstate mall, mine, regional airport, a Disney World, or a new industrial "park." Not yet anyway. In the meantime, it too has become a full-fledged member of the throwaway economy, and its young people still depart in large numbers for careers elsewhere.

If New Wilmington has so far gotten off lightly, other towns and regions have not. Within a few miles, New Castle and Youngstown are industrial disaster areas. The landfill on the outskirts of my present home-town sells space for garbage from as far away as New York City. In south-ern Ohio, the nuclear processing plant at Fernald has spread radioactive waste over several hundred square miles. The same is true of Maxey Flats, Kentucky; Rocky Flats, Colorado; and Hanford, Washington, all sacri-ficed in the name of "national security." Urban sprawl and decaying downtowns afflict hundreds of other towns and cities throughout the United States. Large chunks of footloose capital ravage other places. In northern Alberta, Canada, Mitsubishi Corporation has invested over $1 billion to build a pulp mill that will impair or destroy an ecosystem along with the indigenous culture. One hundred thousand square kilometers of rain forest will be destroyed to supply Europe with cheap pig iron from the Carajas mine in Brazil (Carley and Christie, 1993, p. 24). The result-ing devastation will not show up in the price of steel in Europe. Nor will the devastation from the other mines, wells, clear-cuts, or feedlots around the world, which supply the insatiable appetite of the industrial economy, be subtracted from calculations of wealth. The annual gross world econ-omy now exceeds $21 trillion, and we are told that this must increase fivefold by the middle of the next century. That same global economy now uses, directly or indirectly, 25% of the earth's net primary productivity. Can that increase fivefold as well?

❖ A World That Takes Its Places Seriously ❖

Custodians of the conventional wisdom believe that economic growth is a good and necessary thing. Growth, in turn, requires capital mobility, free trade, and the willingness to take risks and make sacrifices. For the

sake of growth, whole regions and entire industries may have to be sacrificed as production and employment go elsewhere in search of cheaper
labor and easier access to materials and markets. Such sacrifices are necessary, they say, so that "we" can remain competitive in the global economy and so that the things we buy can be as cheap as profit-maximizing
corporations can make them. Conventional wisdom also holds that
"transnational problems cannot be managed by one country acting
alone" (Haas et al., 1993, p. ix). Proponents of the global point of view
often cite the Montreal Accord and subsequent agreements that phase out
chlorofluorocarbons as proof positive.

The first bit of conventional wisdom denies the importance of place
and environment in favor of global vandalism masquerading as progress.
Its more progressive adherents believe that environmental improvement
itself requires further expansion of the very activities that wreck environments. Devotees of the second piece of conventional wisdom ignore the
political and ecological creativity of place-centered people, wishing us to
believe that the same organizations that have ruined places around the
world can be trusted to save the global environment.

On the contrary, a world that takes both its environment and prosperity seriously over the long run must pay careful attention to the patterns that connect the local and the regional with the global. I do not
believe that global action is unnecessary or unimportant. It is, however,
insufficient and inadequate. Taking places seriously would change what
we think needs to happen at the global level. It does not imply parochi-
alism or narrowness. It does not mean crawling into a hole and pulling
the ground over our heads, or what economists call autarky. While we
have heard for years that we should "think globally and act locally," these
words are still more a slogan than a clear program. The national and the
international are still accorded a disproportionate share of our attention,
and the local not nearly enough. I would like to offer five reasons why
places, the local arena, and what William Blake once called "minute particulars" are globally important.

First, we are inescapably place-centric creatures shaped in important
ways by the localities of our birth and upbringing (Gallagher, 1993; Tuan,
1977). We learn first those things in our immediate surroundings, and
these we soak in consciously and subconsciously through sight, smell,
feel, sound, taste, and perhaps other senses we do not yet understand. Our
preferences, phobias, and behaviors begin in the experience of a place. If
those places are ugly and violent, the behavior of many raised in them will

also be ugly and violent. Children raised in ecologically barren settings, however affluent, are deprived of the sensory stimuli and the kind of imaginative experience that can only come from biological richness. Our preferences for landscapes are often shaped by what was familiar to us early on. There is, in other words, an inescapable correspondence between landscape and "mindscape" and between the quality of our places and the quality of the lives lived in them. In short, we need stable, safe, interesting settings, both rural and urban, in which to flourish as fully human creatures.

Second, the environmental movement has grown out of the efforts of courageous people to preserve and protect particular places: John Muir and Hetch-Hetchy, Marjorie Stoneman Douglas and the Everglades, Horace Kephart and the establishment of the Great Smokies National Park. Virtually all environmental activists, even those whose work is focused on global issues, were shaped early on by a relation to a specific place. What Rachel Carson once called the "sense of wonder" begins in the childhood response to a place that exerts a magical effect on the ecological imagination. And without such experiences, few have ever become ardent and articulate defenders of nature.

Third, as Garrett Hardin argues, problems that occur all over the world are not necessarily global problems, and some truly global problems may be solvable only by lots of local solutions. Potholes in roads, according to Hardin, are a big worldwide problem, but they are not a "global" problem that has a uniform cause and a single solution applicable everywhere (Hardin, 1993, p. 278; Hardin, 1986, pp. 145–163). Any community with the will to do so can solve its pothole problem by itself. This is not true of climate change, which can be averted or minimized only by enforceable international agreements. No community or nation acting alone can avoid climate change. Even so, a great deal of the work necessary to make the transition to a solar-powered world that does not emit heat-trapping gases must be done at the level of households, neighborhoods, and communities.

Fourth, a purely global focus tends to reduce the earth to a set of abstractions that blur what happens to real people in specific settings. An exclusively global focus risks what Alfred North Whitehead once called the "fallacy of misplaced concreteness" in which we mistake our models of reality for reality itself, equivalent, as someone put it, to eating the menu, not the meal. It is a short step from there to ideas of planetary management, which appeals to the industrial urge to control. Indeed, it is aimed mostly at the preservation of industrial economies, albeit with greater efficiency. Planetary managers seek homogenized solutions that work against cultural and ecological diversity. They talk about efficiency but not about sufficiency and the idea of self-limitation (Sachs, 1992, p. 111). When the world and its problems are taken to be abstractions, it becomes easier to overlook the fine grain of social and ecological details for the "big picture"; and it becomes easier for ecology to become just another science in service to planet managers and corporations.

A final reason why the preservation of places is essential to the preservation of the world has to do with the fact that we have not succeeded in making a global economy ecologically sustainable, and I doubt that we will ever be smart enough or wise enough to do it on a global scale. All of the fashionable talk about sustainable development is mostly about how to do more of the same, but with greater efficiency. The most prosperous economies still depend a great deal on the ruination of distant places, peoples, and ecologies. The imbalances of power between large wealthy economies and poor economies virtually asssure that the extraction, processing, and trade in primary products and the disposal of industrial wastes rarely will be done sustainably. Having entered the global cash economy, the poor need cash at any ecological cost, and the buyers will deny responsibility for the long-term results, which are mostly out of sight. As a result, consumers have little or no idea of the full costs of their consumption. Even if the sale of timber, minerals, and food were not ruinous to their places of origin, moving them long distances is. The fossil fuels burned to move goods around the world add to pollution and global warming. The extraction, processing, and transport of fossil fuels is inevitably polluting. And the human results of the global trading economy include the effects of making people dependent on the global cash economy with all that it portends for those formerly operating as self-reliant, subsistence economies. Often it means leaving villages for overcrowded shantytowns on the outskirts of cities. It means growing for export markets while people nearby go hungry. It means undermining economic and ecological arrangements that worked well enough over long periods of time to join the world economy. It means Coca-Cola, automobiles, cigarettes, television, and the decay of old and venerable ways. The rush to join the industrial economy in the late years of the twentieth century is a little like coming on board the Titanic just after icebergs are spotted dead ahead. In both instances, celebrations should be somewhat muted.

❖ Implications ❖

The idea that place is important to our larger prospects comes as good news and bad news. On the positive side, it means that some problems that appear to be unsolvable in a global context may be solvable on a local scale if we are prepared to do so. The bad news is that much of western history has conspired to make our places invisible and therefore inacessible to us. In contrast to "dis-placed" people who are physically removed from their homes but who retain the idea of place and home, we have become "de-placed" people, mental refugees, homeless wherever we are. We no longer have a deep concept of place as a repository of meaning, history, livelihood, healing, recreation, and sacred memory and as a source of materials, energy, food, and collective action. For our economics, history, politics, and sciences, places have become just the intersection of two lines on a map, suitable for speculation, profiteering, another mall, another factory. So many of the abstract concepts that have shaped the modern world, such as economies of scale, invisible hands, the commodification of land and labor, the conquest of nature, quantification of virtually everything, and the search for general laws, have rendered the idea of place impotent and the idea of people being competent in their places an anachronism. This, in turn, is reinforced by our experience of the world. The velocity of modern travel has damaged our ability to be at home anywhere. We are increasingly indoor people whose sense of place is indoor space and whose minds are increasingly shaped by electronic stimuli. But what would it mean to take our places seriously?

THE IDEA OF PLACE

First, it would mean restoring the idea of place in our minds by reordering educational priorities. It is commonly believed, however, that the role of education is only to equip young people for work in the new global economy in which trillions of dollars of capital roam the earth in search of the highest rate of return. Those equipped to serve this economy, whom Robert Reich (1991) calls "symbolic analysts," earn their keep by "simplify[ing] reality into abstract images that can be rearranged, juggled, experimented with, communicated to other specialists, and then, eventually, transformed back into reality" (pp. 177–179). Symbolic analysts "rarely come into direct contact with the ultimate beneficiaries of their work"; rather, they mostly sit before computer terminals—examining words and numbers, moving them, altering them, trying out new words and numbers, formulating and testing hypotheses, designing or strategizing. They also spend long hours in meetings or on the telephone, and even longer hours in jet planes and hotels—advising, making presentations, giving briefings, doing deals. (Reich, 1991, p. 179)

Symbolic analysts seem to be a morally anemic bunch whose services "do not necessarily improve society," a fact that does not seem to matter to them, perhaps because they are too busy "mov[ing] from project to project . . . from one software problem to another, to another movie script, another advertising campaign, another financial restructuring" (pp. 185, 237). They are, in Reich's words, "America's fortunate citizens," perhaps 20% of the total population, but they are increasingly disconnected from any interaction with or sense of responsibility for the other four fifths (p. 250). People educated to be symbolic analysts neither have loyalty to the long-term human prospect nor are prepared by intellect or affection to improve any place. And they are sure signs of the failure of the schools and colleges that presumed to educate them but failed to tell them what an education is for on a planet with a biosphere.

The world does not need more rootless symbolic analysts. It needs instead hundreds of thousands of young people equipped with the vision, moral stamina, and intellectual depth necessary to rebuild neighborhoods, towns, and communities around the planet. The kind of education presently available will not help them much. They will need to be students of their places and competent to become, in Wes Jackson's words, "native to their places." They will need to know a great deal about new fields of knowledge, such as restoration ecology, conservation biology, ecological engineering, and sustainable forestry and agriculture. They will need a more honest economics that enables them to account for all of the costs of economic–ecological transactions. They will need to master the skills necessary to make the transition to a solar-powered economy. Who will teach them these things?

ECONOMIES OF PLACE

Taking places seriously means learning how to build local prosperity without ruining some other place. It will require a revolution in economic thinking that challenges long held dogmas about growth, capital mobility, the global economy, the nature of wealth, and the wealth of nature. My views about capital mobility and related subjects were influenced, no doubt, by growing up near a now derelict industrial city, a monument of sorts to mobile capital and failed ideas. Even the prosperous city of my memory, however, was an ecological disaster. On both counts, could it have been otherwise? What would "place-focused economies" look like (Kemmis, 1990, p. 107)?

Historian Calvin Martin (1992) argued that the root of the problem dates back to the dawn of the neolithic age and to the "gnawing fear that the earth does not truly take care of us, of our kind . . . that the world is not truly congenial to sapient Homo" (p. 123). Perhaps this is why most indigenous cultures had no word for scarcity and why we, on the other hand, are so haunted by it. Long ago, out of fear and faithlessness, we broke our ancient covenant with the earth. I believe that this is profoundly true. But we need not go so far back in time for workable ideas. Political scientist John Friedmann (1987) argued that in more recent times

> we have been seduced into becoming secret accomplices in our own evisceration as active citizens. Two centuries after the battle cries of Liberty, Fraternity, and Justice, we remain as obedient as ever to a corporate state that is largely deaf to the genuine needs of people. And we have forfeited our identity as 'producers' who are collectively responsible for our lives. (p. 347)

What can be done? While believing that "the general movement of the last six hundred years toward greater global interdependency is not likely to be reversed," Friedmann argued for "the selective de-linking of territorial communities from the market economy" and "the recovery of political community" (pp. 385–387). This work can only be done, as he put it, "within local communities, neighborhoods, and the household."

But communities everywhere are now vulnerable to the migration of capital in search of higher rates of return. In the case of Youngstown, after the purchase of Youngstown Sheet and Tube by the Lykes Corporation and eventually the LTV Corporation, its profits were used to support corporate investments elsewhere (Lynd, 1982). This money should have been used for maintenance and reinvestment in plant and equipment. Eventually the business failed, taking with it many other businesses. The decision to divert profits out of the community was made by people who did not live in Youngstown and had no stake or interest in it. Their decision had little to do with the productivity of the business and everything to do with shortsightedness and greed.

From this and all too many other cases like it, we can conclude that one requisite of resilient local economies is, as Daniel Kemmis (1990) stated,

> the capacity and the will to keep some locally generated capital from leaving the region and to invest that capital creatively and effectively in the regional economy. (p. 103)

This in turn means selectively challenging the "supremacy of the national market" where that restricts the capacity to build strong regional economies. It also means confronting what economist Thomas Michael Power (1988) called a "narrow, market-oriented, quantitative definition of economics" in favor of one that gives priority to cultural, aesthetic, and ecological quality (p. 3). Economic quality, according to Power, is not synonymous with economic growth. The choice between growth or stagnation is, in Power's view, a false one that "leaves communities to choose between a disruptive explosion of commercial activity, which primarily benefits outsiders, while degrading values very important to residents and being left in the dust and decay of economic decline" (p. 174). There are alternative ways to develop that do not sell off the qualities that make particular communities desirable in the first place. Among these, Power proposed "import substitution" whereby local needs are increasingly met by local resources, not by imported goods and services. Energy efficiency, for example, can displace expensive imports of petroleum, fuel oil, electricity, and natural gas. Dollars not exported out of the community then circulate within the local economy, creating a "multiplier effect" by stimulating local jobs and investment.

Power, like Jane Jacobs in her 1984 book *Cities and the Wealth of Nations*, argued for development

> built around enterprising individuals and groups seeing a local opportunity and improvising, adapting, and substituting. Initially, these efforts start on a small scale and usually aim to serve a local market. (p. 186)

This approach stands in clear contrast to the standard model of economic development whereby communities attempt to lure outside industry and capital by lowering local taxes and regulations and providing free services, all of which lower the quality of the community.

The development of place-focused economies requires questioning old economic dogmas. The theory of free trade, for example, originated in an agrarian world in which state boundaries were relatively impermeable and capital flows stopped at national frontiers (Daly, 1993; Daly and Cobb, 1989, pp. 209–235; Morris, 1990). These conditions no longer hold. Goods, services, and capital now wash around the world, dissolving national boundaries and sovereignty. Labor (i.e., people) and communities, however, are not so mobile. Workers in the developed world are forced to compete with cheap labor elsewhere, with the result of a sharp decline in workers' income (Batra, 1993). For previously prosperous communities, free trade means economic decline and the accompanying social decay now evident throughout much of the United States.

In place of free trade, World Bank economist Herman Daly and theologian John Cobb recommend "balanced trade" that limits capital mobility and restricts the amount that a nation can borrow by importing more than it exports (Daly and Cobb, 1989, p. 231). To restore competitiveness where it has been lost, they recommend enforcing national laws designed to prevent economic concentration (p. 291). To build resilient regional economies, they recommend enabling communities to bid for the purchase of local industries against outside buyers. To the argument that international capital is necessary for the development of third and fourth world economies, they respond that

> we have come, as have many others, to the painful conclusion that very little of First World development effort in the Third World, and even less of business investment, has been actually beneficial to the majority of the Third World's people. . . . For the most part the Third World would have been better off without international investment and aid [which] destroyed the self-sufficiency of nations and rendered masses of their formerly self-reliant people unable to care for themselves. (pp. 289–90)

Daly and Cobb believe that economies should serve communities rather than elusive and mythical goals of economic growth.

Why does the idea that economies ought to support communities sound so utopian? The answer, I think, has to do with how fully we have accepted the radical inversion of purposes by which society is shaped to fit the economy instead of the economy being tailored to fit the society. Human needs are increasingly secondary to those of the abstractions of markets and growth. People need, among other things, healthy food, shelter, clothing, good work to do, friends, music, poetry, good books, a vital civic culture, animals, and wildness. But we are increasingly offered fantasy for reality, junk for quality, convenience for self-reliance, consumption for community, and stuff rather than spirit. Business spends $120 billion each year to convince us that this is good, while virtually nothing is spent informing us what other alternatives we have or what we have lost in the process. Our economy has not, on the whole, fostered largeness of heart or spirit. It has not satisfied the human need for meaning. It is neither sustainable nor sustaining.

THE POLITICS OF PLACE

Taking the environment seriously means rethinking how our politics and civic life fit the places we inhabit. It makes sense, in Daniel Kemmis's (1990) words, "to begin with the place, with a sense of what it is, and then try to imagine a way of being public which would fit the place" (p. 41). I do not think it is a coincidence that voter apathy has reached near epidemic proportions at the same time that our sense of place has withered and community-scaled economies have disintegrated. As with the economy, we have surrendered control of large parts of our lives to distant powers.

Rebuilding place-focused politics will require revitalizing the idea of citizenship rooted in the local community. Democracy, as John Dewey (1954) observed, "must begin at home, and its home is the neighborly community" (p. 213). But neighborly communities have been eviscerated by the physical imposition of freeways, shopping malls, the commercial strip, and mind-numbing sprawl. The idea of the neighborly community has receded from our minds as the centralization of power and wealth has advanced. But neither vital communities nor democracy are compatible with economic and political centralization, from either the right or the left.

We need an ecological concept of citizenship rooted in the understanding that activities that erode soils, waste resources, pollute, destroy biological diversity, and degrade the beauty and integrity of landscapes are forms of theft from the commonwealth as surely as is bank robbery. Ecological vandalism undermines future prosperity and democracy alike. For too long we have tried to deal with resource abuse from the top down and have pitifully little to show for our efforts and money. The problem, as Aldo Leopold (1991) noted, is that for conservation to become "real and important" it must "grow from the bottom up" (p. 300). It must, in other words, become fundamental to the day-to-day lives of millions of people, not just to those few professional resource managers working in public agencies.

An ecologically literate people, engaged in and by its place, will discover ways to conserve resources. Like citizens in Osage, Iowa, they will learn how to implement energy-efficiency programs that save thousands of dollars per household. They will discover ways to save farms through "community supported agriculture," where people pay farmers directly for a portion of their produce. They will limit absentee ownership of farmland and enable young farmers to buy farms. They will find the means to save historic and ecologically important landscapes. They will develop procedures to accommodate environmentalists and loggers, as did the residents of Missoula, Montana. They may even discover, as did residents of the Mondragon area of Spain or the state of Kerala in India, how to successfully address larger issues of equitable development (Whyte and Whyte, 1988; Franke and Chasin, 1991).

We are not without models and ideas, but we lack the vision of politics as something other than a game of winners and losers fought out by factions with irreconcilable private interests. The idea that politics is little more than the pursuit of self-interest is embedded in American political tradition at least from the time James Madison wrote Federalist Paper 10. It is an idea, however, that tends to breed the very behavior it purports only to describe. In the words of political scientist Steven Kelman (1988), "Design your institution to assume self-interest, then and you may get more self-interest. And the more self-interest you get, the more draconian the institutions must become to prevent the generation of bad policies" (p. 51). Kelman proposed that institutions be designed not merely to restrain the unbridled pursuit of self-interest but to promote "public spirited behavior" in which "people see government as an appropriate forum for the display of the concern for others." The norm of public spiritedness also changes how people define their self-interest. This is, I believe, what Vaclav Havel (1992) meant when he described "genuine politics" as "a matter of serving those around us: serving the community, and serving those who will come after us" (p. 6). The roots of genuine politics are moral, originating in the belief that what we do matters deeply and is recorded "somewhere above us."

Is it utopian to believe that our politics can rise to public spiritedness and genuine service? I think not. Evidence shows that we are in fact considerably more public spirited than we have been led to believe, not always and everywhere to be sure, but more often than a cynical reading of human behavior would show (Kelman, 1988, p. 43, notes 38–41). On the other hand, it is utopian to believe that the politics of narrow self-interest will enable us to avert the catastrophes on the horizon that can be forestalled only by foresight and collective action.

❖ Conclusion ❖

Western civilization irrupted on the earth like a fever, causing, in historian Frederick Turner's (1980) words, "a crucial, profound estrangement of the inhabitants from their habitat." We have become, Turner continued, "a rootless, restless people with a culture of superhighways precluding rest and a furious penchant for tearing up last year's improvements in a ceaseless search for some gaudy ultimate" (p. 5). European explorers arrived in the "new world" spiritually unprepared for the encounter with the place, its animals, and its peoples. American settlers' discontent spread to native peoples who were caught in the way. None were able to resist either the firepower or the seductions of technology.

More than just a symbol of a diseased spiritual state, that fever is now palpably evident in the rising temperature of the earth itself. A world that takes its environment seriously must come to terms with the roots of its problems, beginning with the place called home. This is not a simple-minded return to a mythical past but a patient and disciplined effort to learn, and in some ways, to relearn the arts of inhabitation. These will differ from place to place, reflecting various cultures, values, and ecologies. They will, however, share a common sense of rootedness in a particular locality.

We are caught in the paradox that we cannot save the world without saving particular places. But neither can we save our places without national and global policies that limit predatory capital and that allow people to build resilient economies, to conserve cultural and biological diversity, and to preserve ecological integrities. Without waiting for national governments to act, there is a lot that can be done to equip people to find their place and dig in.

SOURCES

Batra, R. 1993. *The Myth of Free Trade*. New York: Scribners.

Berry, W. 1987. *Home Economics*. San Francisco: North Point Press.

Carley, M., and Christie, I. 1993. *Managing Sustainable Development*. Minneapolis: University of Minnesota Press.

Daly, H. 1993, November. The Perils of Free Trade. *Scientific American*, pp. 50–57.

Daly, H., and Cobb, J. 1989. *For the Common Good*. Boston: Beacon Press.

Dewey, J. 1954. *The Public and Its Problems*. Chicago: Swallow Press.

Franke, R., and Chasin, B. 1991. *Kerala: Radical Reform as Development in an Indian State*. San Francisco: Institute for Food and Development Policy.

Friedmann, J. 1987. *Planning in the Public Domain*. Princeton: Princeton University Press.

Gallagher, W. 1993. *The Power of Place*. New York: Poseidon Press.

Haas, P., et al., eds. 1993. *Institutions for the Earth*. Cambridge: MIT Press.

Hardin, G. 1986. *Filters Against Folly*. New York: Viking.

Hardin, G. 1993. *Living Within Limits: Ecology, Economics, and Population Taboos*. New York: Oxford University Press.

Havel, V. 1992. *Summer Meditations*. New York: Knopf.

Hiss, T. 1990. *The Experience of Place*. New York: Random House.

Kelman, S. 1988. Why Public Ideas Matter. In R. Reich, ed., *The Power of Public Ideas*. Cambridge: Harvard University Press.

Kemmis, D. 1990. *Community and the Politics of Place*. Norman: University of Oklahoma Press.

Leopold, A. 1991. *The River of the Mother of God and Other Essays by Aldo Leopold*. S. Flader and J. B. Callicott, eds. Madison: University of Wisconsin Press. (Original work published 1941.)

Lynd, S. 1982. *Fight Against Shutdowns*. San Pedro, CA: Singlejack Books.

Martin, C. 1992. *In the Spirit of the Earth*. Baltimore: Johns Hopkins University Press.

Morris, D. 1990, September/October. Free Trade: The Great Destroyer. *The Ecologist*, 20, pp. 190–195.

Orr, D. 1992. *Ecological Literacy*. Albany: State University of New York Press.

Power, T. M. 1988. *The Economic Pursuit of Quality*. Armonk, NY: M. E. Sharpe.

Reich. R. 1991. *The Work of Nations*. New York: Knopf.

Sachs, W. (ed.) 1992. *The Development Dictionary*. London: ZED Books.

Snyder, G. 1974. *Turtle Island*. New York: New Directions.

Tuan, Y. F. 1977. *Space and Place: Their Perspective of Experience*. Minneapolis: University of Minnesota Press.

Turner, F. 1980. *Beyond Geography: The Western Spirit Against the Wilderness*. New York: Viking.

Whyte, W., and Whyte, K. 1988. *Making Mondragon*. Ithaca: Cornell University Press.

❖

CHAPTER TWENTY-TWO

Prices and the Life Exchanged: Costs of the U.S. Food System

T HE COST of a thing," Thoreau once wrote, "is the amount of what I will call life which is required to be exchanged for it, immediately or in the long run." Thoreau knew what some have yet to discover: the difference between price and cost. Prices, what we pay at the checkout counter, are specific and countable. Costs, on the other hand, include (1) things of value that cannot be measured in numbers; (2) things that could be measured but that we choose to ignore; and (3) the loss of things that we did not know to be important until they were gone. Americans, we are proudly informed, pay only 15% of their disposable income on food, compared with the 23.8% that Europeans pay (National Research Council, 1989, pp. 34–35). But this figure clearly does not represent the true costs of supplying food. Another of my favorite economists, Will Rogers, once said that "It ain't what we don't know that gives us trouble, it's what we know that ain't right." What "we know that ain't right" about the price of food is the source of a great deal of trouble with worse yet to come. The prices we pay for food do not reflect the life we exchange for it or that which we will subsequently forfeit. This is so, in large measure, because life—biotic resources and the health of rural communities essential to a healthy agriculture and culture—is not included in our present accounting system, which instead tends to regard these "factors of production" as if they are as replaceable as worn-out machines.

The practice of ignoring the difference between price and true cost is the stuff out of which historians write epitaphs for whole civilizations. The difference between price and cost is also a matter of honesty and fairness between those who benefit and those who, sooner or later, are required to pay. One effect of not paying full costs is that we fool ourselves into thinking that we are much richer than we really are. Prices that do not "tell the truth," in Amory Lovins's words, eventually lure us (or our children) toward bankruptcy. But the truth that needs to be told cannot be spoken only or even primarily in the language and with the numbers of economics alone. It must be told in the language of ecology, culture, and politics as well.

❖ The True Costs of Food ❖

What are the true costs of the U.S. food system? The first and most discussed are costs resulting from damage to natural systems that accompanies industrial food production. Average soil erosion rates in the United States are estimated to be 7.1 tons per acre per year, 14 times faster than the rates at which soil is created. Cornell University scientist David Pimentel (1990) has estimated that soil erosion and associated water run-off cost the United States $44 billion annually (p. 8). If one considers water overdrafts, land subsidence, soil salinization, and public subsidies for western water, waste of both surface and groundwater costs billions of dollars more than we pay. Pesticides, for which farmers spend $4 billion annually, are estimated to cause $2 to $4 billion in health and environmental damages, including an estimated 20,000 cases of pesticide-caused cancer each year (Pimentel, 1990, p. 11; National Academy of Sciences, 1987). Five billion livestock in the United States produce some 41.8 billion tons of manure each year, half of which is wasted, becoming a source of pollution for groundwater and streams and a source of methane, a powerful greenhouse gas. Agriculture has become progressively more dependent on fossil fuels at an uncalculated cost to the environment from the extraction, processing, transport, and combustion. Food packaging is another source of environmental costs. One third of the solid waste stream is food packaging. David Pimentel has estimated that the total unpriced costs of the U.S. food system fall between $150 and $200 billion dollars per year. A recent study by economists at the World Resources Institute similarly showed that "where everything relevant [is counted], the traditional accounting method's $80-per-acre profit becomes a $26-per-acre loss" (Faeth et al., 1991, p. vi).

A second unpriced cost of the U.S. food system is the loss of farms and rural communities. U.S. Census Bureau data show that America is now overwhelmingly an urban nation, with 77.5% living in metropolitan areas and 50.2% living in large metropolitan areas (*The New York Times*, February 21, 1991). The number of farms in the United States has dropped from 6.5 million in the late 1930s to less than 2 million in 1990. The number of full-time, owner-operated farms is considerably smaller still. Each farm failure means a loss of three to five rural jobs. The loss of six farms means one failed rural business (Strange, 1990, p. 7). Agricultural communities lost nearly 6% of their population between 1969 and 1986 and now have higher poverty rates (17%) than do urban areas (9%). The farm population of 4.6 million (less than 2% of the total population) is now so low that the U.S. Census Bureau will no longer keep separate records on it (*The New York Times*, October 10, 1993). Farms and farming communities are dying, and the U.S. food system is increasingly dominated by "super farms," which are roughly to farming what WalMart is to retailing. What do these things mean?

If one counts only farm "productivity," the gross tonnage of food generated, the answer is "not much." By broader criteria, however, the answer is radically different. Cornell University horticulturist Liberty Hyde Bailey (1980) once questioned

> whether the race can permanently endure urban life, or whether it must be constantly renewed from the vitalities in the rear. (p. 27)

According to Bailey, rural communities

> beget men and women [who are] serious and steady and know the value of every hour and of every coin that they earn; and whenever they are properly trained these folk recognize the holiness of the earth. (p. 29)

Barring a serious national commitment to foster reruralization, the experiment to which Bailey referred is now nearly complete. We are, by and large, an urban people without much vitality in the rear. Rural areas continue to send their young to the cities, and their economies are thoroughly colonized by outside forces, not the least of which is television. We are not, however, more civilized or happier. To the contrary, statistics on virtually every kind of crime, social pathology, and insanity suggest that we are moving in the opposite direction: toward a kind of insensate high-tech barbarism.

Bailey, I think, had it right. We are losing, with barely a whimper, the cultural diversity—special skills, products, cuisine, human qualities, and traditions—that made small towns and rural areas distinctive. Is this just nostalgia? Jacquetta Hawkes (1951), in her classic book *A Land*, put it this way:

> It would be sentimental blindness of another kind to ignore the significance of its [distinctive rural cultures' and economies'] achievement—the unfaltering fitness and beauty of everything men made from the land they had inherited. (p. 144)

And we are losing a choice that a majority of Americans prefer. Gallup poll data consistently show that half or more of Americans would prefer living in small towns (34%) or on farms (22%) than in urban areas (*The New York Times*, September 11, 1990, p. A12).

Along with the loss of a vital rural culture, we are losing the ecological basis on which a rural culture must depend. Rural America is under assault from those who see it as only a place to dump urban refuse and toxic substances, to "recreate," and to speculate (Fritsch, 1989). Acid rain and climate change will only accelerate the destruction.

There is a related loss that is even harder to measure: the loss of the sort of intelligence about the land that once resulted from the close contact with soils, animals, wildlife, forests, and the seasons fostered by farming and rural living. For all of their environmental destructiveness, farms have been in large part the places where Americans were instructed in the realities of nature. To be sure, the lessons were incompletely learned, and sometimes they were not learned at all. But when they were learned and practiced as a kind of craft work, great intelligence was evident and, I think, great satisfactions resulted from a collective effort "to fit close and even closer" into the land (Sturt, 1984, p. 66).

We no longer see agriculture that way but rather as an act of domination to force ever higher yields from the land. Whatever the gains in "productivity," I think we are a less intelligent society than we otherwise might have been. What Gene Logsdon (1984) called "traditional agriculture" stretched and multiplied the intelligence of those who did it well (pp. 3–18). They had to know a fair amount about a great many things: animal husbandry, soil science, nonchemical ways to control bugs and weeds, crop rotations, wood lot management, timber-frame construction, mechanics, and even the weather. Good farmers were good naturalists who knew their places well and knew how to use them well. They are still the best model we have for what is now called "sustainable farming." Moreover, they had to be, in the main, good neighbors and community members. Some of those who still fit this description have learned how to use the sun to dry crops and heat livestock shelters. A few are relearning how to harness the wind. And a handful are way out in front of the society by learning to power their farms with solar-generated hydrogen fuels (Meadows, 1990).

The kind of intelligence evident in good "traditional farming" is in inverse proportion to the amount of purchased "inputs." It is the result of the kind of mind that is willing to be instructed by a place and all that is part of it. Gretel Ehrlich (1990) has described the process in these words:

> Once we understand where and why life occurs and how to stop destroying it, a mindfulness about everything spreads. The land tells us what it needs and when; we just have to be awake, to listen, and to scrutinize the ground . . . a ranch [or farm] is a teacher. (p. 111)

In traditional farming communities, information is passed from generation to generation and is woven into the culture of the region. It is decidedly not the monopoly of a separate caste of farm "experts," or "researchers," who mostly live in another place and occupy another mindscape. For the loss of farms as places of instruction and as a source of practical and ecological competence, I know of no good substitute.

There is a fourth cost of the U.S. food system that also defies easy economic calculation. This is the increasing concentration of wealth and power as agriculture and food processing and distribution have become big business. One effect of concentration is that land is being priced beyond the means of those who must pay for it by farming it (Davidson, 1990, pp. 13–46). Another is that the foods we eat are the product of industrial processes described by longer and longer labels listing chemical additives and ingredients (Rogoff and Rawlins, 1987). Concentration throughout the food sytsem also means that formerly self-reliant rural communities, consisting of owner-operated farms and local markets, have lost control over their economies. Taxes, land-grant university research agendas, and public policies have combined to favor concentration of ownership, suppliers, banks, processors, speculators, and large-scale corporate farming. Jefferson's dream of rural life is rapidly disappearing and with it, in Wendell Berry's words,

> the idea that as many as possible should share in the ownership of the land and thus be bound to it by economic enterprise, by investment of love and work, by family loyalty, by memory and tradition. (quoted in Bryan and McClaughry, 1989, p. 7)

We do not know whether democracy can long survive without widely dispersed control of rural land and resources, but there are good reasons to think that it cannot.

Fifth, in the list of unaccounted costs of the food system are the costs of future investment and capital depreciation, which well run businesses include in current prices (Strange, 1990, pp. 116–117). Agriculture and the food sector have done no such thing. Instead, both have become increasingly dependent upon oil, which is no longer abundant in the United States and the use of which adds to global warming, ecological devastation, and political insecurity. The costs of the transition to renewable sources of energy are not included in the prices we pay for food (Gever et al., 1986, pp. 177–215). If the energy used throughout the food sector were to be priced at the rate of the cheapest renewable alternative, prices would rise dramatically. Whether because of scarcity, restrictions by exporting nations, or the imposition of carbon taxes to prevent climate change, energy prices will rise in coming decades. Agriculture is unprepared for this transition. Nor is it prepared for what some believe may be a period of climate surprises, which may require the resuscitation of farming east of the 100th meridian, where we have been converting prime farmland into housing tracts and shopping malls for the past 50 years. Finally, future costs would have to include those associated with the discovery or rediscovery of how to farm on current solar income. I do not know what the Amish would charge to teach agronomy departments and extension agents such things, but they have a reputation for paying and charging full costs.

Sixth, and finally, the cost of the U.S. food system should include the damage it does to our health. Americans eat too much and too much of the wrong things. As a result we are unique among the nations of the world in the range and novelty of diet-related ailments, such as heart disease, cancer, diabetes, and tooth decay. Diet-related health problems are a sizable part of the nation's medical bill. No doubt, acolytes for the status quo would point out that sizable economic benefits also accrue from the growing sector of the economy that concerns itself with selling remedies: "health" clubs, diet clinics, vendors of exercise videos and books, plastic surgeons willing to "liposuct" various parts of the anatomy, and jogging equipment suppliers. But one could also call these hidden costs of a mismanaged food system.

❖ Sources of the Problem ❖

History, climate, and natural abundance have conspired to make Americans less attentive to cost than we might have been in a less well-endowed land. There are, however, other reasons why we have paid so little attention to the costs of agriculture. Among these, I am inclined to believe that our manner of thinking about economics is the most important. It is no mere coincidence that the discipline of economics began in the same places (Scotland, England, and America) and in the same period as did the industrial revolution. Economics from its early beginnings was stamped with the industrial mind-set and with those assumptions convenient to industrialization. Foremost among these was the belief that the economy ought to be the central institution of modern life. We became Economies with societies instead of Societies with economies. The economy was no longer effectively restrained by obligations to a larger community (Daly and Cobb, 1989). The industrial stamp on economics was also apparent in the belief in the permanence of expansion. The discipline of economics consequently lacked any concept of appropriate scale or sufficiency. Had economists been more attentive to biologists, such egregious errors might have been avoided. Closely related was the belief that human wants were insatiable and should be liberated from the stigma previously associated with greed, avarice, gluttony, and the other "deadly" sins. These former vices were, accordingly, transmuted into economic virtues essential to the proper functioning of the fully modern economy.

If the model of "economic man" was fundamentally deficient, it was at least good for business and for the business of economics. The technologically developed, industrial nations now routinely assume what no other decent culture risked assuming or justifying: that everyone and everything has its price. The market, accordingly, became the arbiter of matters formerly thought to be appraised differently, including land and labor. The industrial mind-set also left economists with an overriding sense of optimism about the beneficence of technology and its ability to overcome the limits of nature. It is now an article of faith in American culture that technology will rescue us from all sorts of ecological mal-feasance and hubris. Finally, the industrial mind-set is evident in the prac-tice of discounting the future. The mythical Dr. Faustus, as economist George Foy has noted, was the first economist to practice discounting, assigning to his soul a 5% discount rate over the 20 years of his contrac-tual association with Mephistopheles. The case he made for the trans-action is, with due allowance for its archaic language, that still made by many economists. By discounting the future value of farm and rural lands to present value, the practice has contributed in no small way to the destruction of the independent family farm and farm communities.

A second reason why costs of the food system have been ignored has to do with the organization of agricultural research and education. The Morrill Act of 1862 (which created land-grant institutions), the Hatch Act of 1887 (establishing agricultural experiment stations), and the Smith-Lever Act of 1914 (the extension service) were intended to improve the lives and livelihoods of rural people by establishing a "permanent agriculture" fostered by local institutions and undergirded by "liberal and practical education." But land-grant universities, in Wendell Berry's (1977) words, "reduced 'liberal and practical' to 'practical,' and then for 'practical' they substituted 'specialized'" (p. 147). The result was to con-vert agriculture from a broadly conceived enterprise with technical aspects and based on a solid agrarian philosophy and moral foundation into a series of technical specializations.

The problem, however, goes deeper. The Hightower Report of 1972, for example, concluded that "Land grant college research is science for sale [and] ultimately it is corrupt of purpose" (Hightower, 1978, p. 85). Many observers believe this, in the main, still to be true. Martin Kenney in his 1986 study of university-industry relations, concluded that "the university is bought and parceled out," and is therefore incapable of con-ducting an unbiased debate on issues such as biotechnology (p. 246). The "subservience" described by Kenney has also precluded unbiased debate within many land-grant universities about increasing farm scale, chemical inputs, farm diversity, organic agriculture, and rural communities. As the mission of land-grant universities became more and more closely identi-fied with the interests of agribusiness, the chemical industry, food engi-neers, the processors, conglomerates, and banks, questions about the full costs of conventional agriculture have been mostly asked outside the U.S. Department of Agriculture (USDA)–land-grant university complex in small nonprofit institutions, such as The Center for Rural Affairs, the Land Stewardship Project, the Land Institute, the Institute for Food and Development Policy, and the Institute for Alternative Agriculture, and by irate citizens.

A third reason why we have ignored full costs has to do with the agricultural policies that defer ecological and social debts to a future electorate. This is not an abstract impersonal process but a failure of political leadership and of particular presidents, senators, congress representatives, and secretaries of agriculture who have failed to concern themselves with long-term costs of the present food system. Instead they have followed the path of least resistance, which means a policy of cheap food and cheap energy for which we will pay dearly in the long run. The loss of family farms, the decay of rural communities, pesticide contamination in groundwater, and the loss of topsoil have happened for reasons that can be found in federal tax laws, commodity programs, USDA-funded university research agendas, farm credit programs, and most recently the North American Free Trade Agreement and the larger free trade agenda contained in the General Agreement on Tariffs and Trade (GATT). The failure to properly reward good farming practices that conserve soil and biological diversity and the failure to support rural communities and rural livelihoods are a product of systematic neglect and studied ignorance of things rural, biological, ecological, and long-term. It is an intellectual failure, a moral failure, a failure of leadership, and a failure of our collective vision.

Finally, the tendency to ignore full costs of the food system must be placed in its larger cultural context. From that perspective, it can be seen not as an aberration but as part of a larger pattern evident in our failure to develop a coherent, long-term energy policy, deal with nuclear wastes, confront the challenge and threat of global warming, eliminate the national debt, deal with the savings and loan crisis, and build decent, sustainable cities. From one perspective, Americans have always been the "people of plenty" in David Potter's phrase, shaped by abundance and consequently inclined to be wasteful. But the tendency to ignore full costs of our actions has, I believe, become worse in the past two decades. Both political parties in this era deserve the harsh judgment of future historians. But so too does the public that elected them and tolerated shortsighted mismanagement of the nation's public estate and its future commonwealth. The failure to distinguish cost and price is part of our national mind-set, a way of thinking that we must now confront.

❖ Toward an Honest Food System ❖

It is always easier to describe a problem than it is to offer sensible and workable solutions to it. This is certainly true of the discrepancy between cost and price in the U.S. food system, which is part of a larger pattern of deeply ingrained values, behavior, and policy. But there is another pattern in American history that we might dust off and put to work. I am referring to the minority tradition represented by Thomas Jefferson, Henry David Thoreau, George Perkins Marsh, Liberty Hyde Bailey, F. H. King, J. I. Rodale, E. H. Faulkner, Russell Lord, Paul Sears, Rachel Carson, Wendell Berry, Wes Jackson, Marty Strange, and many others. This second tradition is a minority refrain throughout our history, but for good reasons, it has never died out. It has been, on the whole, more honest about the costs of the industrial economy than its boosters and benefactors have been. On fundamentals, the big questions of agriculture, food, and national policy, it has been mostly right, and I think there are encouraging signs that it is gaining ground. Drawing from this newer version of agrarianism, I conclude this chapter with four suggestions about the outlines of a food system that pays its full costs.

First, we need an accounting system that includes all of the costs of consumption. Development of an "ecological economics" is a hopeful step in this direction. As stated by Robert Costanza et al. (1991),

> Ecological economics differs from both conventional economics and conventional ecology in terms of the breadth of its perception of the problem, and the importance it attaches to environment-economy interactions. It takes this wider and longer view in terms of space, time and the parts of the system to be studied. (p. 3)

In contrast to conventional economics, ecological economics does not assume that the biosphere is unlimited. Nor does it assume the centrality of human wants or that these should necessarily take precedence over the stability and integrity of the natural systems on which the fulfillment of real needs and wants ultimately depends. It does not make imprudent assumptions about technological progress or the ability of technology to compensate for human stupidity or ignorance. Nor does it assume that technology can be adequately substituted for the loss of "natural capital," such as fertile soils, clean water, abundant wildlife, healthy forests, burgeoning wetlands, intact ozone layers, and stable climate. It does not assume that nature is only a fund that can be depleted at will and without penalty. It does not assume that economic growth is everywhere and at all times a good thing. To the contrary, ecological economics makes a distinction between growth and development and between optimum and maximum. Ecological economics does not discount the future in the manner of conventional economics. Nor does it confuse honest accounting with complex, elegant abstractions far removed from life and lived experience. In short, ecological economics regards the economy as a partial means to higher ends and not as its own end and the study of things economic as "life science," not as the study of greed efficiently practiced (Daly, 1980, pp. 238–252).

Second, we need better farm and food policies that encourage decisions in accord with long-term well-being and that require full-cost accounting. Such policies are simple to describe in the abstract: They will reward ecologically sound agriculture and penalize that which is destructive. The latter, I believe, results in large part from excessive scale of land and machinery and the separation of ownership from management. Destructive agriculture also results inadvertently from federal commodity programs that still encourage farmers to ignore resource costs. Economists at the World Resources Institute showed that "farmers still [after the 1990 Farm Bill] have strong financial incentives to plant just a few crops and use energy-intensive chemical means of fertility maintenance and pest control" (Faeth et al., 1991). These economists concluded that it is possible to "lower the resource costs of U.S. farming, while raising agricultural productivity and lowering the fiscal burden of supporting farm incomes" (p. 29). From a different perspective, Marty Strange (1988) has proposed a variety of policy changes, which include (1) ending subsidies to capital through tax and credit policies; (2) mandatory controls on production; (3) interventions in the land market to equalize opportunity; (4) closer regulation of industries selling farm inputs; and (5) redirection of research to benefit smaller scale farms and reduce adverse impacts (pp. 254–290). While disagreements will undoubtedly occur about specifics, we know enough to use the policy tools available [regulation, pricing, taxes, fees, "feebates," permits, subsidies, and things we have not thought of yet] to achieve a dependable food supply and a sustainable prosperity that is fair and ecologically durable. The heart of the problem, I think, is not one of knowledge or even accurate accounting, but one of political will.

This brings me, third, to matters having to do with our desire to do what is right over the long term. The problem of food and agriculture cannot be reduced to prices and economics alone. Too many of the core assumptions of economics, such as the belief in perpetual growth and the rationality of self-interest over community interests or those of the larger land community, will work against long-term protection of land and rural communities. If economic "efficiency" is the standard, sooner or later someone will win the Nobel prize in economics by showing that one vast, centrally located (presumably in Nebraska) farm can feed the entire nation most efficiently. Another perhaps will win it by showing that it is even more efficient to do away with that farm and food altogether in favor of genetically engineering our capacity for direct photosynthesis. I know of no good reason to place our trust so squarely, so absolutely, and so blindly in the grip of a rationality ultimately so narrow. The only answer is to subordinate a lower rationality to a higher order of rationality.

The prospects for a sustainable agriculture will in the end depend on a larger movement away from the consumer economy toward an economy that supplements efficiency with sufficiency and refuses to place any price whatsoever on priceless things. As Alan Durning (1991) stated, this is an economy that returns

> to the ancient order of family, community, good work, and good life; to a reverence for excellence of skilled handiwork; to a true materialism that does not just care about things but cares *for* them; to communities worth spending a lifetime in. (p. 169)

The recovery of moral and civic virtue Durning proposes is no quick fix. It is rather a long process, perhaps requiring centuries to undo what the industrial economy has done to the land and to us in the past 150 years.

Such an undertaking is nowhere on the national agenda at present. The leadership of this country is mostly in the hands of those proclaiming themselves to be practical and realistic. But the present time, in Lewis Mumford's (1973) words, "is one of those periods when only the dreamers are practical men. By the same token, the so-called practical men have become makers and perpetuators of nightmares" (p. 415). Is the dream of a sustainable economy and healthy cities surrounded by a prosperous countryside only utopian? I think not. Is it utopian to believe that prices ought to include all costs? To the contrary, these are the only practical and realistic alternatives we have.

SOURCES

Bailey, L. H. 1980. *The Holy Earth*. Ithaca: New York State College of Agriculture. (Original work published 1915.)

Berry, W. 1977. *The Unsettling of America*. San Francisco: Sierra Club Books.

Bryan, F., and McClaughry, J. 1989. *The Vermont Papers*. Post Mills, VT: Chelsea Green Publishing Co.

Costanza, R., et al. 1991. Goals, Agenda, and Policy Recommendations for Ecological Economics. In R. Costanza et al., eds., *Ecological Economics*. New York: Columbia University Press.

Daly, H. 1980. On Economics as a Life Science. In H. Daly, ed., *Economics, Ecology, Ethics*. San Francisco: W. H. Freeman.

Daly, H., and Cobb, J. 1989. *For the Common Good*. Boston: Beacon Press.

Davidson, O. G. 1990. *Broken Heartland*. New York: Anchor Books.

Durning, A. 1991. Asking How Much Is Enough. In L. Brown et al., eds., *State of the World: 1991*. New York: Norton.

Ehrlich, G. 1990. Growing Lean, Clean Beef. In R. Clark, ed., *Our Sustainable Table*. San Francisco: North Point Press.

Faeth, P., et al. 1991. *Paying the Farm Bill*. Washington, DC: World Resources Institute.

Fritsch, A. 1989. *Communities at Risk*. Washington, DC: Renew America.

Gever, J., et al. 1986. *Beyond Oil*. Cambridge: Ballinger.

Hawkes, J. 1951. *A Land*. New York: Random House.

Hightower, J. 1978. *Hard Times, Hard Tomatoes*. Cambridge: Shenkman.

Kenney, M. 1986. *Biotechnology: The University-Industrial Complex*. New Haven: Yale University Press.

Logsdon, G. 1984. The Importance of Traditional Farming Practices for a Sustainable Modern Agriculture. In W. Jackson et al., eds., *Meeting the Expectations of the Land*. San Francisco: North Point Press.

Meadows, D. 1990, September 29. Remarkable Energy Savings are Possible. *Valley News*.

Mumford, L. 1973. *The Condition of Man*. New York: Harcourt Brace Jovanovich.

National Academy of Sciences. 1987. *Regulating Pesticides in Food*. Washington, DC: National Academy Press.

National Research Council. 1989. *Alternative Agriculture*. Washington, DC: National Academy Press.

Pimentel, D. 1990. Environmental and Social Implications of Waste in U.S. Agriculture. *Journal of Agricultural Ethics*.

Rogoff, M., and Rawlins, S. 1987, December. Food Security. *Bioscience*, 37, 11.

Strange, M. 1988. *Family Farming*. Lincoln: University of Nebraska Press.

Strange, M. 1990. *Rural Economic Development and Sustainable Agriculture*. Walthill, NE: Center for Rural Affairs.

Sturt, G. 1984. *The Wheelwright's Shop*. New York: Cambridge University Press.

CHAPTER TWENTY-THREE

Refugees or Homecomers? Conjectures About the Future of Rural America

Long before 2030 the trend toward ever larger cities and an increasing ratio of urban-to-rural dwellers is likely to have reversed.

— L. BROWN ET AL.

AMERICA is an overwhelmingly urban and suburban society and is becoming even more so. In 1950 almost half of Americans still lived in rural areas. By 1990, however, the number was less than one in four (22.9%), and only 1.9% Americans lived on farms. (*The New York Times*, September 11, 1990, p. A12). The great forces that have driven urbanization—technological change, population growth, economic growth, and centralization—are widely believed to be permanent features of all modern and postmodern societies. Most of those remaining in rural areas work in extractive industries, or they are poor, retired, or just too stubborn to give up. The urban vision has always depended on the belief that rural life could not compete with the comforts, convenience, affluence, and excitement of city life. The corollaries are (a) that a prosperous and democratic culture does not require a stable and prosperous rural foundation; (b) that a healthy culture does not need, as Wendell Berry once put it, a stream of farm-born and farm-bred young people; (c) that we are smart enough to provision megalopolitan areas with food, water, energy, materials, public safety, transport, employment, and entertainment, haul off the waste, and do all of these in perpetuity; (d) that urban and suburban life can satisfy our deepest human needs; and (e) that we will never change our minds. The idea that Americans might ever return in large numbers to rural areas—whether by choice or otherwise—is, therefore, an affront to both the implicit belief that historical trends are unidirectional and to an ideology that holds that cities, in contrast to everything else, have no maximum size beyond which they decay or collapse.

Yet looking ahead, say a century or more from now, can we see any plausible reasons to think that such limits exist? Are there reasons to believe that urbanization will peak and that large numbers of people, whether by choice or otherwise, will return to smaller cities, towns, villages, and farms? Is it possible that our descendants will regard the age of sprawling, formless cities as a mistake? Is it possible, too, that they might wonder how we could have been so blind to factors leading to large-scale changes that to them will appear to have been obvious all along? What seemingly small factors, now overlooked, might cause such a reversal? Questions like these are familiar to scientists who study chaos and the behavior of nonlinear, complex systems that are highly sensitive to small changes. Is history, urban history in particular, such a system?

Given the stakes, I believe that it would be foolish not to entertain the thought and the implications that flow from it. To do so, however, requires a considerable effort to overcome the spell of trends and big numbers, the power of a pervasive consensus, and the sheer visual impact of vast urban areas. The mind recoils from any prospect that these might someday and in some measure be abandoned and derelict places. But having grown up in and around rust belt cities, I have also grown accustomed to seeing such sights. And I have become convinced that the costs for failing to exercise foresight are rising sharply.

❖ The Logic of Large Numbers ❖

What factors, then, might radically alter the urban prospect? I think that there are at least five, four of which entail trauma to one degree or another and which give reason for the fifth, which is a well thought-out, democratically conceived plan that would allow those wishing to live in rural areas to do so. Public opinion polls have consistently shown that roughly 50% to 60% of Americans would prefer to live on farms and in small towns, were it economically feasible to do so (Speare and White, 1992, p. 94). I am less interested, however, in the exact percentages than in the possibility of reinhabiting rural areas in a way that is ecologically sustainable and socially just and democratic and that enhances both rural and urban prospects.

ENERGETICS

It became possible to herd people into large cities only after it became technologically feasible to concentrate the energy resources necessary to provide for them. Fossil fuels, which could be burned at a rate of our choosing, changed the scale and the nature of the city. Coal, which replaced wood, powered the industrial revolution and made it possible to move materials and food long distances. The railroad broke the dependence of the city on its hinterland. Cheap oil and the automobile did the rest. Without cheap energy cities could not have sprawled as they now do, nor could they have been provisioned beyond a certain scale, considerably smaller than the size of our 100 largest cities. For perspective, the ancient city or that of the Middle Ages depended largely on the surrounding region for food, materials, fuel, and water. That region was not much larger than the radius of the journey that could be made by ox cart in several days.

However, we are approaching the end of the time in which we can burn fossil fuels cheaply and with ecological impunity. Oil production peaked in the United States around 1970 and has been in decline ever since (Tugwell, 1988, p. 141). Worldwide oil production is expected to peak and begin its long descent globally between 2010 and 2020. M. King Hubbert (1969), who foresaw this transition as early as the 1950s, described it in these words:

> It now appears that the period of rapid population and industrial growth that has prevailed during the last few centuries, instead of being the normal order of things and capable of continuance into the indefinite future, is actually one of the most abnormal phases of human history. It represents only a brief transitional episode between two very much longer periods, each characterized by rates of change so slow as to be regarded essentially as a period of non-growth. It is paradoxical that although the forthcoming period of non-growth poses no insuperable physical or biological problems, it will entail a fundamental revision of those aspects of our current economic and social thinking which stem from the assumption that the growth rates which have characterized this temporary period can be permanent. (p. 239)

There are alternatives to fossil fuels, such as hydrogen, but they will not be cheap. And, on the present scale of energy use, there are no inexpensive and easy transitions to different fuels. The end of the era of cheap energy is a fundamental economic turning point. Throughout most of the twentieth century, we were able to substitute low-cost energy for labor, and doing so vastly increased labor "productivity." One farmer using fossil energy for fuels and fertility, for example, can feed some 75 people. Such numbers, however, conceal the fact that it takes 11 calories of fossil energy to put 1 calorie on the consumer's plate.[1] Remove the fossil fuel subsidy, and the standards for efficiency change as well. Subsistence farmers, operating at far lower "efficiency" levels, invest 1 calorie to produce 50 calories of food (Steinhart and Steinhart, 1974).

The end of the fossil fuel era means that we face, from Vaclav Smil's (1991) perspective, a

> transition from fossil-fueled energetics to a civilization running, again, on instantaneous solar flows . . . but such a transition will be neither fast nor easy. (p. 311)

Even if this transition occurs without major trauma, it probably will mean smaller cities and more dispersed populations due to the high cost of capturing and storing abundant but diffuse sunlight. Energy prices will rise along with those of everything that requires energy for transportation and processing, including food. If the transition from fossil fuels to sunlight is made badly, without foresight or planning, which is plausible, it could well mean supply interruptions, shortages of one kind or another, and economic collapse, all of which would hit urban areas particularly hard. For example, food is transported, on average, 1,300 miles from where it is grown or produced to where it is eaten. Sharply rising energy costs mean both higher production costs and higher transportation costs. With overstocked supermarkets, it is hard for us to imagine the prospect of food shortages, but they could happen. We permitted agriculture to decline in areas where both rainfall and population density are high and subsidized it in arid regions. The result is a brittle food system that works well only if energy is cheap and ecological costs are ignored.

ECOLOGICAL RESILIENCE

A second factor that will affect the balance between urban and rural areas is the possibility of global ecological disruption in the coming century. Because of the emission of heat-trapping gases, for instance, many scientists believe that we are now "committed" to an approximately 1.5°C to 4.5°C increase in the average temperature of the earth sometime in the middle decades of the next century and more thereafter (Intergovernmental Panel on Climate Change, 1990). This is a rate of climate change one thousand or more times faster than any we have experienced before, leading to temperatures hotter than any *Homo sapiens* has experienced. If not arrested, climate change threatens to raise sea levels, accelerate the rate of species extinctions, spread diseases, increase pollution, and increase the likelihood of drought, heat waves, and superstorms. Rising sea levels threaten to inundate cities on low-lying coasts and also make them more prone to storm damage. Because of the "heat island" effect, heat waves will be more severe in the cities than in the countryside. The effects of climate change, in short, will permeate virtually every facet of life. Those who believe that we can adapt to such changes assume that the costs of adaptation will be within our means and that there will be no surprises, sudden shifts, or serious unanticipated effects along the way (see Chapter 12).

The magnitude of the change ahead is daunting. Global emissions of carbon dioxide now total approximately 8.5 billion tons from all sources. The total necessary to stabilize the climate is believed to be 3–4 billion tons per year or less. The United States emits between 1.2 and 1.6 billion tons of carbon dioxide per year, or about 5 tons of carbon dioxide per person per year. Given a world population of 8 billion by 2030, simple arithmetic indicates that, on average, the global per capita carbon dioxide emissions must fall to 0.5 ton per person or less (Intergovernmental Panel on Climate Changes, 1990).

Other large-scale ecological disruptions, such as the rapid loss of species, the loss of topsoil, acid rain, toxic pollution, and population growth, will further diminish the urban prospect. The past 10,000 years of climate stability and ecological abundance in which *Homo sapiens* emerged may be drawing to an end. If so, the economic and social arrangements that can only work under conditions of relative ecological stability will be sorely tested. Among such arrangements, large urban areas may be the most vulnerable. They cannot provision themselves. They do not "work" unless they can exploit surpluses produced in distant places. Ecological instability and climate change will reduce those surpluses and disrupt the long chains of supply and exploitation on which cities are utterly dependent.

HEALTH

A third factor that may diminish the urban prospect in coming decades has to do with the vulnerability of concentrated populations to new and virulent diseases (McMichael, 1993). Not long ago it was widely believed that medical science was about to eliminate all infectious disease. But the race between microbes and humankind is far from over. Old diseases like tuberculosis and hepatitis have returned in more deadly strains resistant to present antibiotics. New diseases like AIDS are spreading rapidly. On the horizon are worse viruses, like Ebola, that can be spread as an aerosol with a sneeze or a cough (Preston, 1992, p. 62). The emergence and spread of these viruses is often linked to the destruction of tropical habitats in which they had once been contained (Chivian et al., 1993, p. 216; Preston, 1992).

Not only are concentrated populations more vulnerable to such diseases, but industrial efforts to provision them with food, energy, and materials have adversely affected health by spreading toxic substances and chemicals virtually everywhere. The human immune system is under assault from factors ranging from increased ultraviolet radiation to toxic substances in the home and workplace. For reasons said to be unknown, sperm counts in young men worldwide have dropped by some 50% since 1938 (Chivian et al., 1993, p. 218). Human breast milk now often exceeds U.S. Environmental Protection Agency (EPA) standards for toxicity. The typical human corpse now contains too many heavy metals and toxic substances to meet EPA standards for landfills (Hawken, 1993, p. 71). Cancer rates worldwide have increased over the past four decades. Rapid climate change will compound health problems by allowing diseases to enter new regions unchecked.

Armed with modern science, we like to think ourselves invulnerable to circumstances such as those of the fourteenth century when one third of the population between India and Iceland died of the bubonic plague (Tuchman, 1978). Perhaps we are. It would be unwise, however, to place all of our faith in that assumption. "There are," in A. J. McMichael's (1993) words, "ominous signs that we now face a heightened constellation of [health] problems in the world's hypertrophied cities" (p. 168). While medical science is advancing in some respects, in others it has hardly moved at all. Where people once died of smallpox and malaria, they now die of coronary disease, obesity, hypertension, cancer, and AIDS. The environment is still the primary determinant of human health, and the decline in the vital signs of the earth will adversely affect human mortality and longevity no matter how good our medical science is.

PROBLEMS OF SCALE AND COMPLEXITY

Fourth, there is good reason to believe that urban areas have grown beyond our ability to manage them well. The usual litany of problems— crime, corruption, drugs, homelessness, the costs of maintaining infrastructure, air and water pollution, waste disposal, and public safety—beset virtually all large U.S. cities and, increasingly, their suburbs as well. Such problems have many sources, including mismanagement, lack of foresight, racism, and capital flight, which can occur to some degree at any scale. But beyond a certain size, the limits of information, organizational flexibility, human neighborliness in mass society, and accountability in a global economy make urban problems more likely and less solvable. Cities, in other words, have limits of scale beyond which they become less manageable, humane, democratic, and livable places. Undoubtedly these limits vary, but they probably do not exceed the size of what is now described as a "small" city.

These four factors—the end of the fossil fuel era, the decline in ecological resilience, emerging threats to human health, and the difficulties of managing complex urban areas—will feed on each other. In other words, they are strongly interactive, and they will create feedback loops that will intensify problems that in isolation might have been solvable.

CHOICE

The cumulative, long-term effects of higher energy costs, ecological overshoot, the vulnerability of concentrated populations to new diseases, and urban breakdown will only tip the balance against further urbanization. They will not force people to return to rural areas, and certainly not in a way that solves either rural or urban problems. The examples of dozens of other overcrowded, poverty-ridden, disease-infested cities testify to human endurance and inertia in the face of utterly inhuman circumstances. But even if they want to leave, urban people may be trapped by other factors. The costs of buying land and moving are beyond the means of many. Jobs in the rural areas are often hard to find. Returning to the farm is not likely for people who have forgotten how to farm and who no longer have family living on farms. Moreover, rural life will continue to be disparaged by image makers, media moguls, spin doctors, and all of those who milk dependent and gullible people. If this were not enough, it is becoming harder to find the country. All too often what one finds is the ruin that accompanies colonization of once rural places: dumps, rural slums, freeways, theme parks, second home developments, mobile home "parks," resorts, and uncontrolled sprawl. It is easier to buy rural real estate than it is to discover anything resembling true country.

There is nonetheless a fifth possibility: To avoid a deepening crisis, we could choose to reinhabit rural areas in an orderly, knowledgeable, and sustainable fashion. Such a response would require an unprecedented degree of foresight and ecological imagination. The alternative, however, is that under stress people will flee to rural areas as refugees, compounding rural problems without solving urban ones. And the cities left behind will fall further into ruin and disrepair. Were we to choose to reinhabit rural landscapes, what would be required to do so sustainably, with foresight and ecological imagination? What does it mean for Americans to turn their stay in this land from an encampment into a durable civilization?

❖ An Honest History ❖

CONQUEST

Before setting out to reinhabit America, we should know what went wrong the first time, by which I mean an ecologically honest account of our past. The fact is that Europeans came to the new world armed with ideas, philosophies, religion, and dreams of wealth that did not fit gracefully and permanently into the place they called America. It is not necessary to romanticize Native Americans to know that our ancestors did not, by and large, care to know anything about the native cultures whose ideas did fit the geographies of America. As a result, with the exception of a few pioneering naturalists, our ancestors often failed to comprehend the ecology of the land or the virtues of the people who had learned to live sustainably on it. Instead, they saw only empty real estate and savages. Neither European culture nor the Christian religion prepared them to be humble, cautious, inquiring, or peaceful. The civilization they imposed on the "new world" reflected European ideas and the arrogance of the conqueror, not the native experience with this land. We were, as Frederick Turner (1980) expressed it, "brought into contact with [our] psychic and spiritual past . . . a contact for which [we] were utterly unprepared" (p. 95).

We have always had difficulty seeing the American land clearly. Puritans tended to see America as a howling wasteland. Later, others saw it as a paradise that would alter its rainfall to suit our agricultural ambitions. In a remarkable assertion about the facts of will on weather, for example, Charles Dana Wilbur wrote the following in 1881:

> In this miracle of progress, the plow was the avant courier . . . not
> by any magic or enchantment, not by incantations or offerings, but,
> instead, in the sweat of his face, toiling with his hands, man can per-
> suade the heavens to yield their treasures of dew and rain upon the
> land he has chosen for his dwelling place. It is indeed a grand con-
> sent, or, rather, concert of forces—the human energy or toil, the vital
> seed, and the polished raindrop that never fails to fall in answer to
> the imploring power or prayer of labor. (quoted in Smith, 1961, p.
> 211)

To my knowledge will has no greater effect on the biophysical realities of
America now than it did in 1881. Yet many persist in the belief that the
lands, forests, mineral wealth, waters, and air of America can be made to
fuel endless economic growth. The ecological record of these ideas is now
indelibly etched on the face of the land. In its totality, it is a record of
environmental vandalism and plundered resources on a scale with no historical precedent.

THE MYTH OF PROGRESS: We failed to write an honest history in
part, I believe, because Europeans came to the new world inclined to
believe too much in the doctrine of progress: that the present is better than
the past and that the future will be better still. The concept of progress
has led us to denigrate native cultures, peasant communities, villages, and
small towns, which is to say, our roots. We have been unable to comprehend what Peter Laslett (1971) called *The World We Have Lost*. Modern
people tend to assume that anything prior to the year before last is hopelessly antiquated and therefore does not need to be understood or preserved. Those who believe otherwise are deemed preservationists and
naive romantics. There is a contrary view, however. Native cultures certainly had their share of imperfections and folly, but as Philip Slater
(1974) stated,

> All the errors and follies of magic, religion, and mystical traditions
> are outweighed by one great wisdom they contain—the awareness of
> humanity's organic embeddedness in a complex natural system. And
> all the brilliant, sophisticated insights of Western rationalism are set
> at naught by the egregious delusion on which they rest—that of
> human autarchy. (p. 233)

The distinguished anthropologist Robert Redfield (1967) reached broadly
similar conclusions about peasant cultures that possessed

> an intense attachment to native soil; a reverent disposition toward
> habitat and ancestral ways; a restraint on individual self-seeking in
> favor of family and community; a certain suspiciousness, mixed with
> appreciation of town life; a sober and earthy ethic. (p. 78)

Villagers everywhere live in an organic world permeated by life and mystery (Critchfield, 1983). They are superstitious, religious, poor, resilient, and humble. They work with their hands, often cooperatively, and they do not understand city peoples' preoccupation with wealth, privacy, and intellect. Some of these are qualities that we should endeavor not to erase.

Throughout our own history, Americans have been ambivalent about our farm and small town origins. The ambivalence is evident in the writings of Edgar Lee Masters, Carl Sandburg, Vachel Lindsay, and Sherwood Anderson, each of whom came from small towns. Those, like Sandburg and Masters, who disliked the narrowness and hypocrisy of small towns, tended to romanticize the city. Consequently a fair view of small town and rural America is hard to come by, not only because of the size and regional diversity of America but because the experience of small town and rural life occurred in the context of a technologically dynamic, capitalist society that undermined tradition and stable communities. We swept across the continent always in search of something better somewhere else. In Wallace Stegner's (1987) words,

> Our migratoriness has hindered us from becoming a people of communities and traditions, especially in the West. It has robbed us of the gods who make places holy. It has cut off individuals and families and communities from memory and the continuum of time. It has left at least some of us with a kind of spiritual pellagra, a deficiency disease, a hungering for the ties of a rich and stable social order. (p. 22)

RURAL LIFE: An honest history, then, will also acknowledge that rural life has often been lonely, dull, and precarious. Rural towns and people have all too often been narrow, bigoted, and violent. While we celebrate the frontiersman and cowboy, rural life, as Walter Prescott Webb (1981) phrased it, "exerted a peculiarly appalling effect on women" (p. 506). An honest accounting of rural America, then, must include the often tragic stories of women, Native Americans, African Americans, and migrant laborers. It must also include the narrowness, small town gossip, and self-righteousness that were often a part of rural communities, along with the genuine kindness, charity, community, and neighborliness they exhibited at their best.

An honest history will also acknowledge the dignity of small farming and rural livelihoods. While public taste regards sexism, ageism, and racism to be offensive, it has yet to discover ruralism, by which I mean the portrayal of rural people as dim-witted, lazy, incompetent, and even slightly dangerous (Logsdon, 1994, pp. 49–56). The urban image of rural people is a cross between the country bumpkin and the loser who couldn't make it in the city where the smart people are. Television programs such as *Hee Haw*, *Dukes of Hazard*, and *Green Acres* reinforce those images among rural people. An honest view of rural life should not be pompous, self-righteous, or unable to poke fun at itself, but it should not acquiesce in having urban people define what it means to be rural, even under the excuse of entertainment.

POLITICS AND POPULISM: An honest history will also acknowledge the effects of corporate centralization, absentee ownership, and inequities that go back to the Colonial era. Rural America has been long dominated by absentee capital, large corporations, and land speculators. But it could have been otherwise. In the election of 1896, Populists attempted to extend democracy to issues of rural land ownership, banking and investment decisions, and the control of large corporations. They failed and with them, "the idea of substantial democratic influence over the structure of the nation's financial system . . . quietly passed out of the American political dialogue" (Goodwyn, 1978, p. 269). Since that time, whole areas of common life have been walled off from public scrutiny and civic dialogue. Now,

> 'The people,' though full of anxiety, do not know what to do politically to make their society less authoritarian. Language is the instrument of thought, and it has proven difficult for people to think about democracy while employing hierarchical terminology. (Goodwyn, 1978, p. 318)

❖Twenty-First-Century Agrarianism ❖

To reinhabit rural America sustainably, we will need better ideas than those typical of the industrial era. In many cases we need to rediscover tried-and-true ways of doing things; in others we need to invent something new. But from whatever source, these ideas must

- work with the ecologies of particular places,
- conserve land and resources over the long run,
- be understandable by people affected by them,
- be affordable, and
- promote democracy and participation.

In the centuries to come, in other words, information must substitute for mass and cheap energy. I propose eight such ideas, each more information-rich than its industrial counterparts. These are not offered in isolation but as parts that must operate together.

THE SUSTAINABLE FARM

The heart of rural America has been the family farm. Of the 6.7 million farms in existence in 1935, less than 2 million have survived, and most of these depend heavily on off-farm income. As family farms gave way to agribusiness, agricultural productivity rose, but so too did environmental damage, energy use, and capital requirements. The inefficiencies of agribusiness, however, are cloaked by public subsidies and dishonest accounting that ignores, among other things, ecological costs (Strange, 1988). Unpriced costs of food in the United States may be as high as $150 billion per year (Pimentel, 1990). We need smaller scale farms, powered by sunshine, that conserve biological diversity, serve local and regional markets with a diverse array of crops and products, and sequester carbon dioxide. They will need to be tailored to the specific conditions of particular places, and they will need, more than ever, to be ecologically resilient, intensively managed, and economically agile. We will need to attract a new generation of young people to agriculture and make such enterprises affordable and profitable for them. And on the whole, young people will need to be better farmers than youths of earlier generations (Montmarquet, 1989, p. 245).

SUSTAINABLE FORESTRY

The industrial age regarded forests mostly for what could be taken from them. Climate and ecological trends ahead indicate that forests will become more important for the carbon they sequester and the diversity of life they shelter. Accordingly, we need new ideas about how to eliminate the 50% of U.S. timber harvest that is waste (Postel and Ryan, 1991, pp. 86–87), how to protect existing forests, and how to regenerate new ones. We need to know how forestry can be done sustainably, not over decades, but over millennia (Aplet et al., 1993). We need better ideas about the big factors that govern the use of forests having to do with absentee corporate ownership and discount rates used by resource economists, which stack the deck against long-term resource husbandry.

LAND AND LAND OWNERSHIP

After John Locke, the modern world regarded land mostly as private property, the value of which equaled the labor necessary to "improve" it. The idea of private property and the idea which grants corporations the rights of individuals are behind much of the sprawl, ugliness, and ecological deterioration of the American landscape. In the century ahead we will need ideas embodied in law and custom that restrict sprawl and abuse while rewarding good land use practices that promote higher density, conserve soils and open spaces, and protect critical habitats and biological diversity. Property as the exclusive domain of individuals and corporations must be amended to protect community rights and those of future generations (Worster, 1993, p. 110). The means to do these things will include old ideas such as land trusts, communal ownership, and the separation of ownership from development rights. While we are at it, it is time to rethink how and why we charter corporations and the reasons of public interest that should cause us to terminate those charters (Hawken, 1993).

VILLAGE AND COMMUNITY

In many places in America, small towns are languishing and have been for quite some time. To accommodate people returning to rural areas in the century ahead, small towns must be revived and made inviting, viable, and accessible places. Rural schools, hospitals, and infrastructure, fallen into decay, must be restored and improved. This will require, however, a spirit of innovation and renewal that is not often characteristic of small communities. But I know of no formula for such a revival beyond pride and love for one's community, plus a dose of imagination for what it might be.

TRANSPORTATION

The end of the fossil fuel era means the end of automobile and truck travel as we have known them. Amidst all of the options for alternative transportation, the best and cheapest is to rebuild the railroad network that for nearly a century bound the country together. We currently spend some

$300 billion each year to subsidize automobiles and trucks (Nadis and MacKenzie, 1993) and another $100 billion or more to subsidize fuels (Romm, 1992). These subsidies should be used instead to rebuild a genuinely national railroad network over, say, 50 years. Railroads once serviced small towns throughout rural America. They will have to once again. This, too, is an old idea that was dismantled, not because it did not work, but because it was systematically subverted in the interest of capital.

THE REGION

The end of the era of cheap energy also means that we need to shorten supply lines for food, materials, and energy whenever possible. The global economy must give way to predominantly regional economies that do not require the energy and environmental costs necessary to process, package, refrigerate, and transport heavy things in bulk over long distances. We must, in short, rediscover the region as an economic and ecological resource. How are we to organize for this transition?

Regional planning has a long and distinguished, if largely ignored, tradition that began with the formation of the Regional Planning Association of America (RPAA) in 1923. Some of the most creative thinking ever done about regions and the intelligent inhabitation of landscapes was done under the auspices of the RPAA by planners and scholars such as Lewis Mumford, Benton MacKaye, and Howard Odum (Friedmann and Weaver, 1980, pp. 29–41). That and more recent work in regional planning provide a framework for integrating rural and urban areas.

ECONOMIES OF PLACE

The conventional theory of rural development deals mostly with how to bring outside capital into rural areas to perpetuate growth. But if capital is mobile enough to come into rural areas to take advantage of cheap labor and resources, it is mobile enough to leave when these can be found still more cheaply elsewhere. The record of development from the outside in is at best mixed. Typically, it provides fewer jobs and at lower wages than initially promised (Lingeman, 1980, pp. 465–470). Imported industries seldom pay their fair share of taxes because of the various incentives and tax abatements that were used to lure them. Often the same lures include exemption from pollution controls, which causes a decline in the quality of the environment, one of the features that makes rural areas attractive in the first place. We need a new model of development recognizing that

> projects that seek to expand employment by attracting new workers into the community do not really develop the community, while projects that improve the range and efficiency of the economic activities of local people do develop the local community. (Daly and Cobb, 1989, p. 134)

In contrast to conventional indexes of development, such as increasing sales, business volume, personal income, number of jobs, and population, economist Thomas Michael Power (1988) proposed that we measure development by the availability of satisfying work, secure access to biological and social necessities, stability, and those things that make communities stimulating, diverse, and vital: qualities that are often undermined by conventional development.

Properly used, outside capital and resources can help rural economies, but they should not be the core of the local economy. That core must be made up of a diversity of local producers and businesses that use local resources and materials to meet local needs, thereby reducing requirements for imported capital, energy, materials, and food to the lowest practical level. A high degree of self-reliance not only multiplies economic opportunities by keeping money in local circulation, it also provides a buffer against outside economic disruptions. In short, economic development must protect rural communities and resources at risk to footloose capital while providing the basis for enduring prosperity.

THE GREEN CITY

It is foolish to think that we can reinhabit rural areas sustainably without also changing the way we inhabit urban areas. Rural prospects mirror those of cities, and one cannot be improved much without improving the other. Many of the same ideas and economic forces that destroyed communities in Appalachia for cheap coal or depleted Kansas prairies and groundwater for cheap food, helped to destroy urban communities in Newark, Youngstown, south Chicago, and in hundreds of other cities. And without serious efforts to redesign urban areas to meet the ecological and human challenges of the twenty-first century, no effort to rebuild rural America will work for long. The urban vision of reruralization is the green city. It is a more ecologically complex city than that of the industrial era (Calthorpe, 1993; Van der Ryn and Calthorpe, 1986). "Greening" cities requires stretching our ecologicial imagination to break down the dichotomy betwen urban and rural and allow rural things—city farms, gardens, forests, wildlife corridors, river parks, and natural areas—into the urban world. It means using ecological technologies to purify wastewater and restore toxic places. It means architecture and urban planning that begin with the natural world and take full advantage of the free services of nature. It means economies built on efficiency, pollution prevention, and renewable energy. It means recognizing limits to the size and geographic sprawl of urban areas. And it means designing cities to fit more closely their surrounding regions.

❖ A Proposal ❖

The probable effects of ecological, climate, and social trends now visible on the horizon indicate that we cannot safely and sanely continue to herd ourselves into cities and suburbs. Neither can we go back in large numbers to rural regions as they are at present. Accordingly, I propose that we create a nationwide effort to build sustainable rural communities and forge a national consensus to carry it out. We have foreign policies, urban policies, economic policies, but virtually no coherent, well thought-out, ecologically grounded, and farsighted policy for rural areas, and certainly none to revitalize rural communities on a sustainable basis. Without such a policy, any future in which substantial numbers of people wish to return to rural areas will be chaotic and bleak. Given the scope of the task and the enormity of the stakes, how might we begin?

One possibility is to create a Presidential Commission to study the matter. There is a long tradition for such a proposal and an equally long shelf of deservedly unread, undigested, and uninspiring reports by Presidential Commissions. Their functions have been mostly cosmetic, to make it appear as if Washington is about to rouse itself to solve some great national problem or other while preserving the status quo that gave rise to the problem in the first place. Presidents, being political animals, are inclined to appoint the politically influential to such Commissions, plus a maverick for balance. It is unlikely, therefore, that a visionary and vigorous national effort to revitalize rural areas would begin in such a desultory fashion.

Another possibility is to assign the task to the land-grant universities created by Congress. In the Morrill Act of 1862, the Hatch Act of 1887, and the Smith-Lever Act of 1914, Congress intended to promote "liberal and practical education" and to provide for "the establishment and maintenance of a permanent and effective agriculture." It has not worked out the way it was intended. The education done in land-grant universities has not been liberal, and the agriculture it fosters is not permanent. Land-grant institutions lack the boldness, vision, flexibility, and intellectual agility required to revitalize themselves, let alone lay the groundwork for revitalizing rural America. Even friends of the land-grant system have come to the conclusion that execution of its existing mandate may require "a new type of organization" (Meyer, 1993, p. 88).

I propose, accordingly, that the process begin by utilizing an existing group of grassroots, nonprofit organizations already working on issues of food, agriculture, and rural revitalization. For the past two decades, the best ideas about sustainable agriculture, rural communities, rural economics, issues of justice, and farm policy have come from organizations such as the Rodale Institute, the Land Institute, the Mississippi Delta Project, the Center for Rural Affairs, the Federation of Southern Cooperatives, the New England Food Association, and the Land Stewardship Project. These and similar organizations are small, flexible, and driven by a level of commitment that far exceeds their budgets. I propose that a group of such organizations in each region of the country be funded for a decade or longer and commissioned to (1) develop detailed proposals for the reruralization of their own regions; (2) organize participation necessary for a genuinely grassroots democratic process; (3) cooperatively develop the ecological, educational, and economic basis for a comprehensive national rural policy; and (4) develop plans to train the leadership in each community and region necessary for the transition ahead. I am proposing, in other words, that the task of revitalizing rural areas begin in rural areas with rural people and rural organizations.

Past efforts to reruralize in the 1930s (Borsodi, 1972) and in the 1970s failed largely because of naivete, the lack of organization, and the lack of a viable rural economy in which those returning to the land might fit. Most, but not all, were beaten individually by economic difficulties. This suggests to me that priority should be given to questions having to do with livelihood, sustainable rural economics, and cooperative endeavors that run counter to the model of the rugged individual.

Given the enormity of the task that lies ahead, a decade of the preparatory work by a dozen or so regional institutes represents only the first step of a long journey that will extend over several centuries. But if that work is done well it will clarify the goals of reruralization and lay the groundwork for the necessary economic, political, legal, and policy changes. Moreover, it will have begun the effort to educate and mobilize the public for the transition to the post–fossil fuel world. It will also have helped to clarify the dimensions of sustainable cities and their ecological relationships to surrounding regions.

❖ Conclusion ❖

E. F. Schumacher (1973) once contrasted homecomers with "people of the forward stampede" (pp. 146–149). The homecomer, comparable to the prodigal son in the Biblical parable, returns after a long binge in shame and penitence. The parable requires both a worthy home to which the prodigal wants to return and the capacity of the prodigal for genuine penitence. Like the prodigal, Americans too have been on a binge, ours fueled by fantasies of power, wealth, and mobility. For us, coming home means restoring ecological and human scale to a civilization that has lost its sense of proportion and purpose. It means regenerating roots in particular places and traditions. But if we do not build a worthy home, what are we building? And if we do not prepare our young people to come home, for what destination and for what destiny do we consign them?

SOURCES

Aplet, H., et al. 1993. *Defining Sustainable Forestry*. Washington, DC: Island Press.

Borsodi, R. 1972. *Flight from the City*. New York: Harper Colophon.

Brown, L., et al. 1990. *State of the World: 1990*. New York: Norton.

Calthorpe, P. 1993. *The Next American Metropolis*. Princeton, N.J.: Princeton Architectural Press.

Chivian, E., et al., eds. 1993. *Critical Condition*. Cambridge: MIT Press.

Critchfield, R. 1983. *Villages*. New York: Anchor Books.

Daly, H., and Cobb, J. 1989. *For the Common Good*. Boston: Beacon Press.

Debeir, J.-C., et al. 1991. *In the Servitude of Power*. London: ZED Books.

Friedmann, J., and Weaver, C. 1980. *Territory and Function*. Berkeley: University of California Press.

Goodwyn, L. 1978. *The Populist Movement*. New York: Oxford University Press.

Hawken, P. 1993. *The Ecology of Commerce*. New York: Harper Business.

Hubbert, M. K. 1969. Energy Resources. In *Resources and Man*. San Francisco: W. H. Freeman.

Intergovernmental Panel on Climate Change. 1990. *Climate Change*. New York: Cambridge University Press.

Laslett, P. 1971. *The World We Have Lost*. New York: Scribners.

Lingeman, R. 1980. *Small Town America*. Boston: Houghton Mifflin.

Logsdon, G. 1994. *At Nature's Pace*. New York: Pantheon.

McMichael, A. J. 1993. *Planetary Overload*. New York: Cambridge University Press.

Meyer, J. 1993. The Stalemate in Food and Agricultural Research, Teaching, and Extension. *Science*, *260*, pp. 881–882.

Montmarquet, J. 1989. *The Idea of Agrarianism*. Moscow: University of Idaho Press.

Nadis, S., and MacKenzie, J. 1993. *Car Trouble*. Boston: Beacon Press.

Pimentel, D. 1990. Environmental and Social Implications of Waste in U.S. Agriculture. *Journal of Environmental Ethics*.

Postel, S., and Ryan, J. 1991. Reforming Forestry. In L. Brown et al., *State of the World: 1991*. New York: Norton.

Power, T. M. 1988. *The Economic Pursuit of Quality*. Armonk, NY: Sharp.

Preston, R. 1992, October 26. Crisis in the Hot Zone. *The New Yorker*, pp. 58–81.

Redfield, R. 1967. *The Little Community/Peasant Society and Culture*. Chicago: University of Chicago Press.

Romm, J. 1992. *The Once and Future Superpower*. New York: Pantheon.

Schumacher, E. F. 1973. *Small is Beautiful*. New York: Harper.

Slater, P. 1974. *Earthwalk*. New York: Bantam Books.

Smil, V. 1991. *General Energetics*. New York: John Wiley & Sons.

Smith, H. 1961. *Virgin Land*. New York: Vintage Books.

Speare, A., and White, M. 1992. Optimal City Size and Population Density for the Twenty-First Century. In L. Grant, ed., *Elephants in the Volkswagen*. New York: W. H. Freeman.

Stegner, W. 1987. *The American West as Living Space*. Ann Arbor: University of Michigan Press.

Steinhart, C., and Steinhart, J. 1974. Energy Use in the U.S. Food System. *Science*, *184*, pp. 307–316.

Strange, M. 1988. *Family Farming*. Lincoln: University of Nebraska Press.

Tuchman, B. 1978. *A Distant Mirror*. New York: Knopf.

Tugwell, F. 1988. *The Energy Crisis and the American Political Economy*. Stanford, CA: Stanford University Press.

Turner, F. 1980. *Beyond Geography*. New York: Viking.

Van der Ryn, S., and Calthorpe, P., eds. 1986. *Sustainable Communities*. San Francisco: Sierra Club Books.

Webb, W. P. 1981. *The Great Plains*. Lincoln: University of Nebraska Press.

Worster, D. 1993. *The Wealth of Nature*. New York: Oxford University Press.

ENDNOTES

1. Even this figure is misleading since 20% of the U.S. labor force is engaged in transporting, processing, and marketing food (Debeir et al., 1991, p. 146).

Conclusion: Earth in Mind

OW ARE minds to be made safe for a planet with a biosphere? One answer is to load students down with more facts and data having to do with the decline of one thing or another. As teachers, educators, and concerned citizens we are obliged to tell the truth as accurately as we see it, which means partially through a glass darkly. But part of the truth cannot be told; it must be felt. It is within us. It would be odd indeed if several million years of evolution had not equipped us for this moment of truth. And the actuality is that, try as we may, we have not and cannot escape the fact that we are, in Stan Rowe's (1993) words, highly evolved deep air mammals. We are of the earth; our flesh is grass. We live in the cycle of birth and death, growth and decay. Our bodies respond to daily rhythms of light and darkness, to the tug of the moon, and to the change of seasons. The salt content of our blood, our genetic similarity to other life forms, and our behavior at every turn give us away. We are shot through with wildness. Call it biophilia (Wilson, 1984) or the ecological unconscious (Roszak, 1992), the earth is inscribed in us, we are of the earth. We have an affinity for nature. What do we do about that simple but overwhelming fact?

The short answer is to face it, but we are still caught up in denial. The civilization we have built causes us to spend 95% of our lives indoors, isolated from nature. "Being born and raised," as Michael Cohen (1993) stated, "bewildered (wilderness-severed) assaults our thinking and our inner nature." We live stress-filled lives full of traffic jams, busyness, noise, artificiality, and substitutes for the real thing. Our culture is riddled with stress and stress-related pathologies: addictions, broken marriages, violence, and greed. More than 70% of our medical problems, costing $250 billion, are believed to be stress related (Cohen, 1993). We are estranged from our sources. We have, in Herman Daly's words, an infinite itch, but we do not know where to scratch.

Were we to confront our creaturehood squarely, how would we propose to educate? The answer, I think, is implied in the root of the word *education, educe,* which means "to draw out." What needs to be drawn out is our affinity for life. That affinity needs opportunities to grow and flourish, it needs to be validated, it needs to be instructed and disciplined, and it needs to be harnessed to the goal of building humane and sustainable societies. Education that builds on our affinity for life would lead to a kind of awakening of possibilities and potentials that lie largely dormant and unused in the industrial-utilitarian mind. Therefore, the task of education, as Dave Foreman stated, is to help us "open our souls to love this glorious, luxuriant, animated planet" (Roszak, 1993). The good news is that our own nature will help us in this process, if we let it.

How will this awakening occur? Scott Momaday (1993) put it this way:

> Once in his life a man . . . ought to give himself up to a particular landscape in his experience, to look at it from as many angles as he can, to wonder about it, to dwell upon it. He ought to imagine that he touches it with his hands at every season and listen to the sounds that are made upon it. He ought to imagine the creatures there and all the faintest motions of the wind. He ought to recollect the glare of noon and all the colors of the dawn and dusk. (p. 83)

SOURCES

Cohen, M. 1993. Integrated Ecology: The Process of Counseling with Nature. *The Humanistic Psychologist,* 21, 3.

Momaday, S. 1993. *The Way to Rainy Mountain.* Albuquerque: University of New Mexico Press. (Original work published 1969.)

Roszak, T. 1992. *Voice of the Earth.* New York: Simon & Schuster.

Roszak, T. 1993. Beyond the Reality Principle. *Sierra* (March/April).

Rowe, S. 1993. Stan Rowe on the Ecological Perspective in a Changing World. *Hastings Bridge,* 1, 1.

Wilson, E. O. 1984. *Biophilia.* Cambridge: Harvard University Press.

INDEX